Advances in Intelligent Systems and Computing

Volume 675

Series editor

Janusz Kacprzyk, Polish Academy of Sciences, Warsaw, Poland
e-mail: kacprzyk@ibspan.waw.pl

About this Series

The series "Advances in Intelligent Systems and Computing" contains publications on theory, applications, and design methods of Intelligent Systems and Intelligent Computing. Virtually all disciplines such as engineering, natural sciences, computer and information science, ICT, economics, business, e-commerce, environment, healthcare, life science are covered. The list of topics spans all the areas of modern intelligent systems and computing.

The publications within "Advances in Intelligent Systems and Computing" are primarily textbooks and proceedings of important conferences, symposia and congresses. They cover significant recent developments in the field, both of a foundational and applicable character. An important characteristic feature of the series is the short publication time and world-wide distribution. This permits a rapid and broad dissemination of research results.

More information about this series at http://www.springer.com/series/11156

Roman Szewczyk · Denis Havlik
Editors

Recent Trends in Control and Sensor Systems in Emergency Management

Editors
Roman Szewczyk
Industrial Research Institute for Automation and Measurements
Warsaw
Poland

Denis Havlik
Department of Safety and Security
AIT Austrian Institute of Technology
Seibersdorf
Austria

ISSN 2194-5357 ISSN 2194-5365 (electronic)
Advances in Intelligent Systems and Computing
ISBN 978-3-319-70451-7 ISBN 978-3-319-70452-4 (eBook)
https://doi.org/10.1007/978-3-319-70452-4

Library of Congress Control Number: 2017957562

Printed on acid-free paper

This Springer imprint is published by Springer Nature
The registered company is Springer International Publishing AG
The registered company address is: Gewerbestrasse 11, 6330 Cham, Switzerland

Foreword

Worldwide, emergency management takes a significant organizational effort, requiring a large amount of planning and forecasting to provide a swift response. Although effective, many existing operating frameworks tend to have shortcomings in a number of areas, including the interconnectivity between emergency operators during the crisis periods. In crisis situations, emergency response teams may often miss available information to take key decisions needed on the ground. This situation may happen, even when appropriate resources are available and the seamless communication is provided.

To improve this situation, the 7th Framework Programme project C2-SENSE: "Interoperability Profiles for Command/Control Systems and Sensor Systems in Emergency Management" financed by the European Commission is seeking to achieve a seamless interoperability across all layers of communication channels in emergency management and crisis response fields.

The aim of C2-SENSE's framework is to provide a proof of concept demonstrating how to improve interoperability by using the Profiles. Within the project, it was elaborated how to develop and also how to validate the conformity of the different systems to the "contract" that is defined in such Profiles. Some of the developed tools are strongly coupled with the Profiles and cannot be replaced by any of the existing systems, as they directly use Profiles as input or as output. Other proposed tools that are less strongly coupled with the Profiles can easily be used outside of the C2-SENSE framework. This way, the pilot application relies mainly on the use of legacy tools for most of the user interaction. In fact, Profiles could even be used to describe and verify the interoperability in an already existing C2 system.

This book presents nine chapters focused on the most important advances of C2-SENSE project. Chapters cover description of analyses and practical advances in physical and sensors layer, data analyses, visualization and management, and system integration. Moreover, certification guidelines and organization interoperability issues are addressed.

Papers are the result of intensive discussion during the Workshop "Command, Control and Sensor Systems in Emergency Management" which was held in

Milano (Italy) from 4 to 6 July 2017. We hope that this book will be the base for further discussion and provide practical guidelines for effective implementation of novel solutions in IT technologies for civil protection applications.

Acknowledgments. This project has received funding from the European Union's Seventh Framework Programme for Research, Technological Development and Demonstration under grant agreement no. 607729.

Contents

Emergency Maps Tool as a Collaborative Instrument for Decision Makers in a Command and Control Environment

Gerald Schimak[1(✉)], Denis Havlik[1], Peter Kutschera[1], Refiz Duro[1], and Mert Gencturk[2]

[1] AIT- Austrian Institute of Technology, Donau-City Straße 1, 1220 Wien, Austria
{gerald.schimak,denis.havlik,peter.kutschera, refiz.duro}@ait.ac.at

[2] SRDC Software Research and Development and Consultancy Corp., Ankara, Turkey
mert@srdc.com.tr

Abstract. Tools for decision making with in a proper collaboration environment are of utmost importance for crises managers as well as decision makers to get continuous and accurate information about the crises situation. One of the main tools in C2-SENSE project is called Emergency Maps Tool (EMT). It has been developed as part of a broader collaboration environment in the C2-SENSE Emergency Interoperability Framework. It is the C2_SENSE instrument for decision making. EMT allows decision makers to set and monitor activities, send and receive event related messages but also to include ad-hoc information from sensors or sensor networks (e.g. water monitoring sensors in case of flooding).

Keywords: Command and Control · Interoperability · Crises- · Disaster and emergency management · Decision making · Map tool · Collaboration

1 Introduction

A proper collaboration environment is of utmost importance for crises managers as well as decision makers to get continuous and accurate information about the crises situation and to manage the available resources on the fly. Interoperability of existing systems, tools, methods and standardized processes are needed to allow effective management of emergencies, crises and disasters. Crises-/Emergency Management and decision making tools have to support and monitor all activities between the involved actors (e.g. authorities, first responders, volunteers, etc.).

In C2-SENSE (Interoperability Profiles for Command/Control Systems and Sensor Systems in Emergency Management, www.c2-sense.eu) a tool, called Emergency Maps Tool (EMT), has been developed as part of a broader collaboration environment in the C2-SENSE Emergency Interoperability Framework. This tool aims to display all relevant resources (e.g. authorities, organization, object of interests (like roads, railways, bridges), messages, alarms, etc.) involved or of special importance, in order to allow proper management of the crises situation. EMT allows decision makers to set and

R. Szewczyk and D. Havlik (eds.), *Recent Trends in Control and Sensor Systems in Emergency Management*, Advances in Intelligent Systems and Computing 675,
https://doi.org/10.1007/978-3-319-70452-4_1

monitor activities, send and receive event related messages but also to include ad-hoc information from sensors or sensor networks.

Backbone of this tool is a data, communication and collaboration model realized in a flexible, configurable and extendable way. That includes a kind of database where the so-called Object of Interests (OOI) are stored, an Enterprise Service Bus (ESB) where all communication between the involved system entities is transported and a manifold of adapters that connects to the ESB.

OOIs can be everything, spanning from metadata of responsible authorities to civil protection departments up to alarms, pure measurement values stemming from sensors, or information/messages about blocked roads as well as number of endangered peoples at a specific location.

2 Collaboration Environment

One of the main issues for managing crisis situations, during or after a natural or man-made disaster event has occurred, as shown in [1, 2], is the lack of interoperability. Different crisis responders and involved organizations usually have their own Command & Control (C2) Systems. Adding to these different procedures, languages, practices and cultural differences increases the complexity of interoperability, especially when a crisis crosses countries and political borders. When the goal is to increase situational awareness through collaboration and joint response, these dispersed systems need to cooperate, communicate, exchange and share data and information. In other words, territorial emergency management requires a cross-organizational, cross-domain, cross-level interoperable collaboration environment between the involved C2-Systems.

Such collaboration environment is developed within the C2-SENSE Emergency Interoperability Framework. It aims to connect all relevant organizations and services that have to cooperate in an emergency or crisis situation as shown in Fig. 1. In a crisis scenario suitable for a framework, several responders and organizations (from now on referred to as actors) attempt to coordinate their efforts with the main goals of better organizing, more effective prioritization of response actions, quicker response time, and more effective resource allocation (i.e. material or personnel) [3].

Other crucial information for the actors on the ground might be provided by sensor networks in flood-related crisis situations (e.g. water level information, road and bridge conditions, flooded areas, etc.). Accurate information and coordination is needed. As shown by [4], the better the coordination becomes, the higher the chances for more effective response and rescue will get.

For situations, as described above, the collaboration environment of the C2-SENSE Emergency Interoperability Framework is highly applicable. Its central part is the Interoperability Knowledge Layer as shown in Figs. 2 and 3 below and in [5]. This layer within the C2-SENSE architecture includes various methods and components to provide a common operational picture of a crises situation, while supporting joint decision making. Moreover, the knowledge interoperability profiles, as described in [6], enable exchange of information and data, and communication between different devices, proprietary or/and non-proprietary alike. In other words, the whole system with these developments becomes fully interoperable. Before going into details of the EMT and OOI we briefly describe the architectural approach and different interoperability levels.

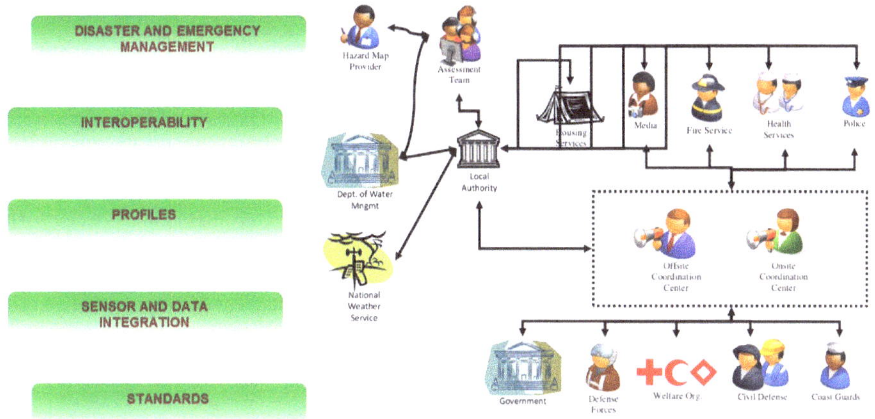

Fig. 1. Collaboration of different actors in an emergency situation through one emergency interoperability framework (represented by solid black line)

3 Architectural Levels

As it can be seen in the interoperability stack shown in Fig. 2, knowledge layer plays a key role in the architecture as it resides in the centre and provides the connection between technical and organizational part. Data coming from systems involved in emergency

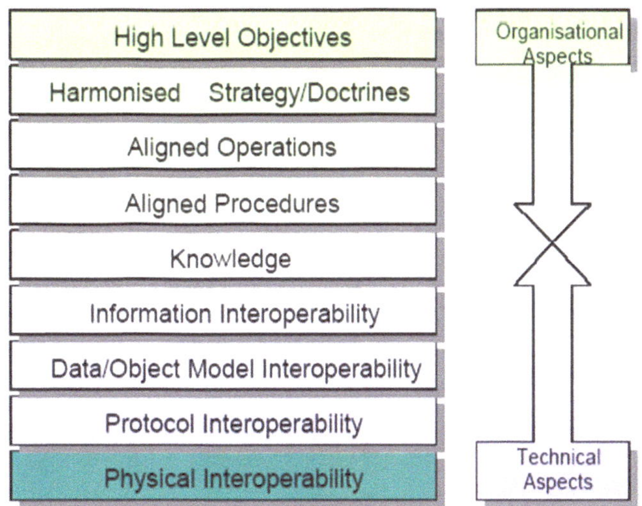

Fig. 2. Interoperability stack

management such as fire department system, hospital information system, sensor networks etc. flows through physical, protocol, data/object model and semantic interoperability layers respectively and reaches the knowledge layer. During this flow, proprietary data is exposed to several transformations so that when it reaches the knowledge layer, it is meaningful and fully complies with the specifications defined in the C2-SENSE Interoperability Profiles [6].

Figure 3 illustrates the overall architecture of C2-SENSE Emergency Interoperability Framework. The quantified specific objectives for each layer of the Interoperability Stack shown above are described in [7] as follows:

- Physical Interoperability Layer: managing the physical connections between the networked applications and devices.
- Protocol Interoperability Layer: managing end-to-end delivery of messages and documents.
- Data/Object Model Interoperability Layer: managing data exchange among the disparate systems through common standard interfaces.
- Semantic Information Interoperability Layer: managing provision of semantic mediation among different but overlapping common standard interfaces.

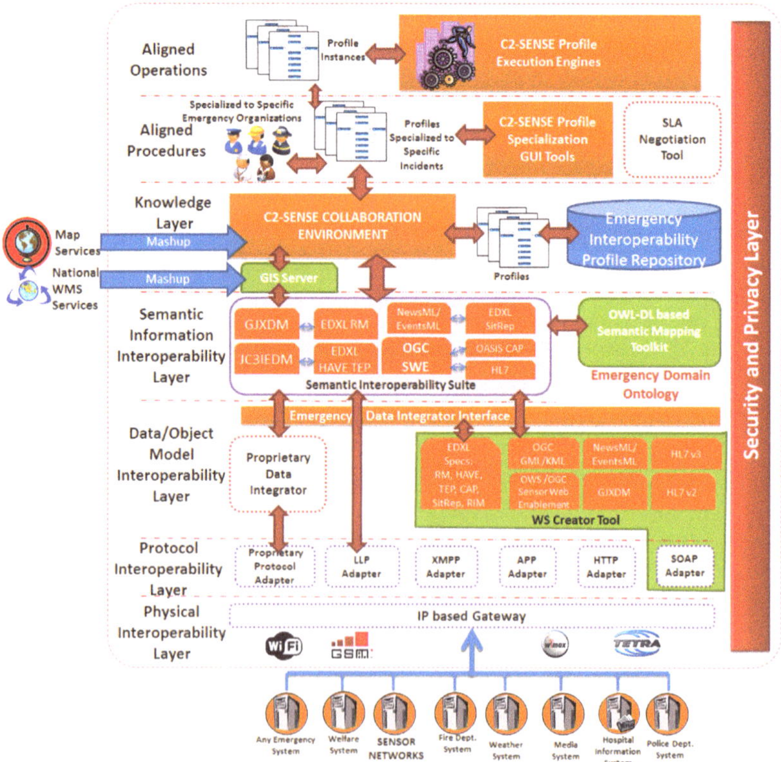

Fig. 3. Overall C2-SENSE architecture

- Knowledge Layer: managing creation of a common operational picture of the crisis situation and having the support of collaboration for joint decision making.
- Aligned Procedures and Operations Layer: managing alignment of emergency partners on their procedures and operations and reaching of an agreement.

In knowledge layer, as shown in Fig. 4, the data is retrieved by collaboration environment consisting of several individual components. Then according to type of data, data is managed by the corresponding component, e.g. sensors are managed by Sensor Management Tool, ambulance positions, blocked roads, sensor measurements are shown in Emergency Maps Tool, alerts are broadcasted by Messaging and Communication Platform etc. As a result, a common operational picture is provided to emergency responders and joint decision making of them is enabled.

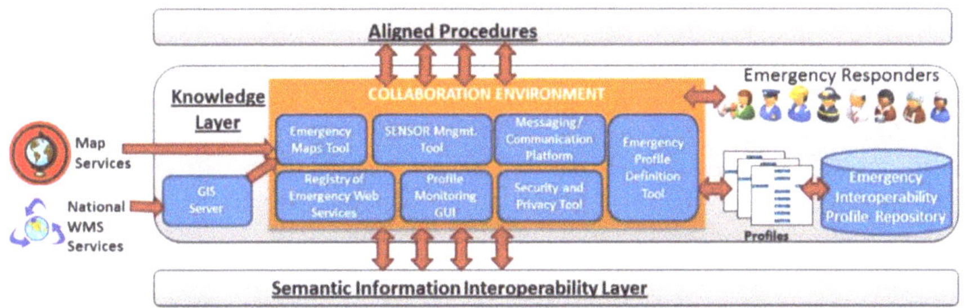

Fig. 4. Knowledge layer architecture

The main components forming the collaboration environment are:

- Emergency Maps Tool, for crisis mapping;
- Sensor Management Tool, for integration and management of different sensors, even ad-hoc ones;
- Messaging/Communication Platform, to send and receive messages (e.g., alarms, reports, etc.);
- Registry of Emergency Web Services, to register web services used in the framework
- Profile Monitoring Tool, to monitor the execution of the implemented emergency process/workflow;
- Profile Definition Tool, to define profiles (e.g., map related profiles, situation analysis profiles)
- Security and Privacy Tool, to protect against security issues;
- Profile Repository, for storing of defined crises/event relevant profiles;
- Object Of Interest Data Repository, the data storage for all objects of interests to be provided within a decision support tool (e.g., the Emergency Maps Tool).

The central part of the collaboration environment is the Emergency Maps Tool aiming to provide a common view of the emergency situation in form of maps of the crises/event area and related situation information. It is the main front end (client side).

With situation information (e.g. features) we mean all the information about objects that are relevant for the visualisation of the emergency situation and a proper decision making for the various users. Therefore when we speak about features we mean also resources (objects of interest) of utmost importance during an event, such as ambulances, infrastructure (e.g., hospitals), or areas of interest such as for example evacuation zones (as shown in Fig. 8).

4 EMT and Its Components

The Emergency Maps Tool aims to display all relevant resources involved or of special importance, in order to allow proper decision making. Resources can be ambulances, buildings related to organizations, collapsed bridges or critical areas. EMT allows decision makers to set and monitor activities, send and receive event related messages, but also to include ad-hoc information from sensors or sensor networks (e.g., water monitoring sensors in case of flooding).

The Web-Browser based Emergency Map Tool is much more than just a map. Beside the interactive map component showing actual information, it combines functionalities to organize information views in different ways to facilitate more effective and quicker decision making. Trends of real-time sensor data can be shown in a graph (Fig. 5), messages can be filtered by type, source or geolocation. Directly from the EMT also messages and commands can be sent out whenever necessary.

A very useful EMT feature, compared with other solutions, is the flexibility and extensibility provided by the used implementation approach. The GUI components are independent of each other and can easily be extended and combined in various ways to fulfil different end user needs.

4.1 EMT GUI Components

Important for the graphical user interface (GUI, see Figs. 5 and 8) is that all the components (i.e., widgets) are interconnected and can interact, thus they are not just stand alone components. For example, an area selected in the Map component can be used to filter messages in the MessageList component.

Further, all components share the same localization features. This includes not only the localization of the GUI itself but also the localization of the data represented. Moreover, this localization is not limited to a selected language, but can also depend on a user domain and interest. This allows user specific data representations in form of icons on the map or in the tables, colouring of messages, and even translation of messages into different user languages (Figs. 7 and 8).

The most relevant EMT GUI components are, a:

- **Map**, which displays a map with interactive OOIs (filtered by type) and overlays (e.g., weather data, flooding maps). Clicking on OOIs opens popup windows, showing latest data related to this OOI) or opens a linegraph to allow a detailed analysis of the data (e.g. latest sensor values).
- **MessageList**, which is a table listing all OOI data. The settings of the list allow to show or hide columns and to apply filters (like filtering by OOI and message type as well as by geo location). Selecting a row can place a marker and/or center the map

to this OOI. Also a more detailed view of the data (e.g. a simple linegraph showing the last days data) can be triggered.
Two types of views for the MessageList are available: (a) Message View: Just the last messages arrived. This can include multiple messages related to one OOI. This just depends on the frequency of the messages. (b) Last Value Table: It provides a more condensed view, as only the last message for each message type and OOI is shown.

- **Linegraph,** which can be used as a GUI component by itself offering a lot of configuration possibilities for data display like colours, time spans and legend.
- **Commands,** which is the place to enter new commands. The Commands component is highly flexible. It allows any message and even sensor data to be sent manually (Fig. 6). A lot of settings are available to tailor the Commands functionality to the user's needs. This includes settings adjusted for a specific domain, but also multilinguality (see Fig. 7). For efficient usage a set of commands can be prepared in advance allowing messages to be sent with one click.
- **Configuration Manager,** which stores the settings of all components. Different configurations for different use cases and/or users can be stored by name, transferred between computers and reloaded later.

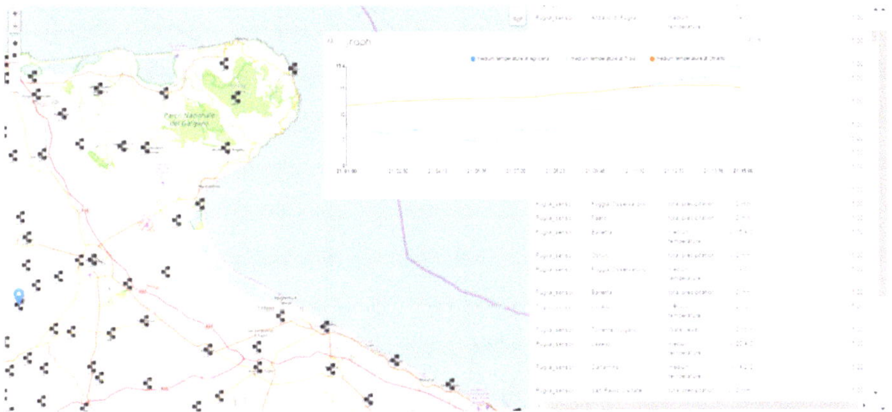

Fig. 5. EMT GUI components: Map, MessageList and Linegraph [3]

Fig. 6. Editing box for messages [3]

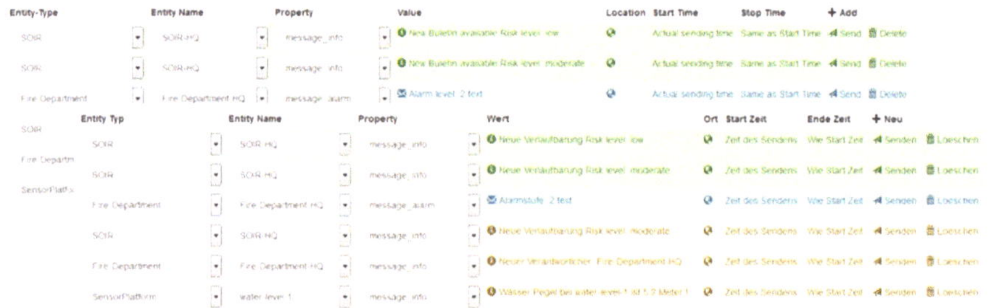

Fig. 7. Commands component (screenshot) – settings for languages. The values themselves (column "Wert") are also translated [3]

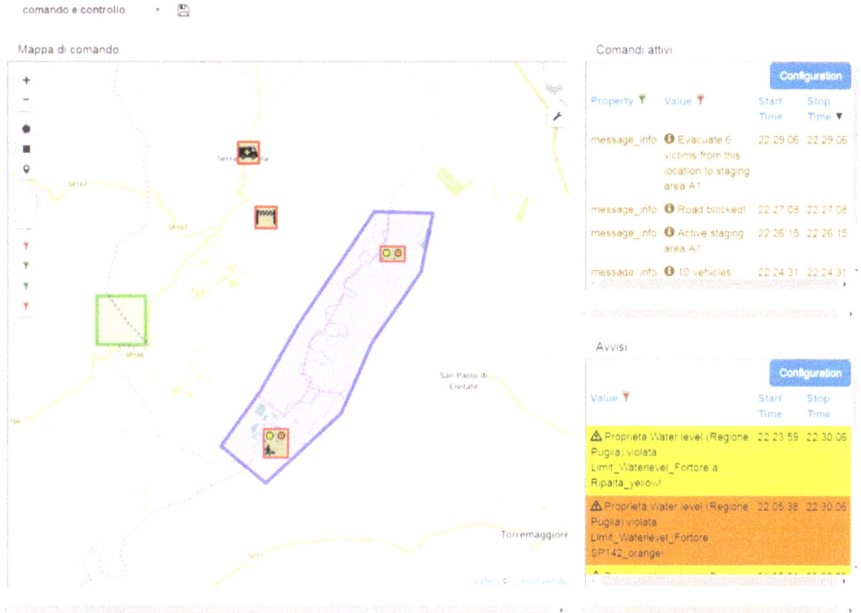

Fig. 8. Multilingual emergency maps tool with a map and two table views, one configured as a command log, the other as a table of active alerts (in a selected area)

5 OOI – Object of Interest Data Repository

5.1 The Architecture and the Purpose of the OOI

The main data repository for the EMT is the "Object Of Interest" repository holding all relevant data to be shown in the EMT. The EMT is, hence, tightly connected to the OOI (see Fig. 10). This guarantees fast response times for the GUI. This is an important feature for crisis responders, as stated in [8]. Additional data sources can be used by the EMT as well. This includes external GIS server providing background maps, or other sources connected over the internet. In the case of a crisis with degraded internet connectivity, however, these external data sources may be unavailable, so all the necessary data and maps should be in the OOI or other local services, e.g., a local service providing the background maps.

Currently, the only interface needed is the REST interface [9] of the OOI. Apache Kafka [10] is used as Enterprise Service Bus (ESB) providing reliable communication between components within C2-SENSE and the stakeholder systems.

The OOI data repository is connected to ESB using a set of listeners. This makes it very easy to add new data types (Fig. 9) from any sources being translated into OOI entries. The OOI-administration interface allows the management of different OOI types. Moreover the OOI data can be used by other C2-SENSE components and it can also provide data for external usage. The existing OGC WMS [11] and OGC WFS [12] based on GeoServer [13] provide OOI data to clients/actors that already have well established tools. OOI data are offered in the form of map layers, the two most relevant ones are: “lastvalues” containing the latest, most actual numeric OOI values. This layer is used for sensor data. “lastmessages” delivers messages as JSON objects. This allows different presentations depending on the user needs but here the translation to different user languages and domain specifics is left to the client.

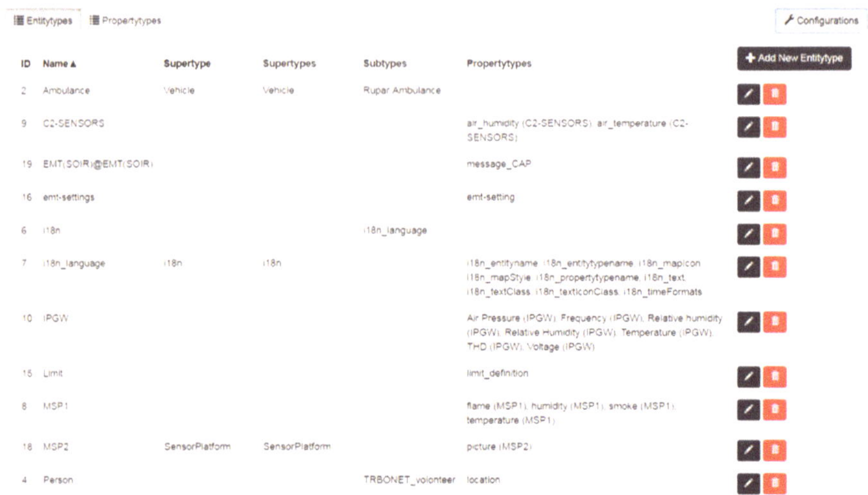

Fig. 9. OOI administration interface, showing some of the currently defined OOI entity types.

5.2 Analyzing the OOI Data

LimitChecker is a component in direct connection to the OOI data repository. It is an important component for meeting some of the demands of a detection and decision support system, as mentioned in [2]. One of the main functionality in the context of the C2-SENSE Emergency Interoperability Framework is performing analysis of data, like the verification of incoming sensor values for detecting and identifying critical situations in the field and producing warning or alarming messages for a crisis manager. For example, during a flooding situation, field sensors measuring water level will stream the data to the OOI repository. These values will then be compared to water level limits already provided and defined by the emergency managers, and in case of exceedance, LimitChecker will produce/initiate an appropriate warning or alarming message (Fig. 11).

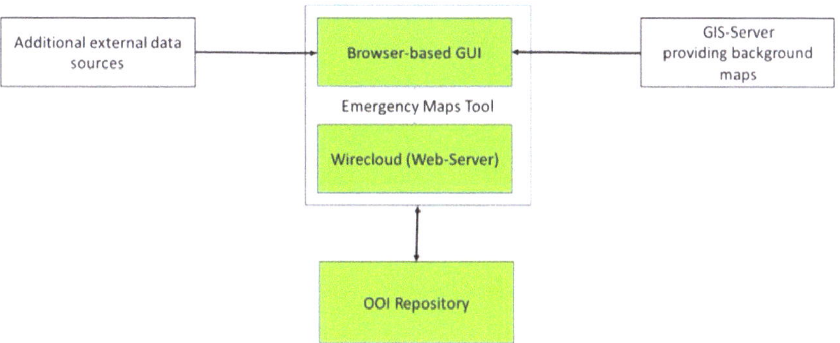

Fig. 10. EMT and OOI overview [3]

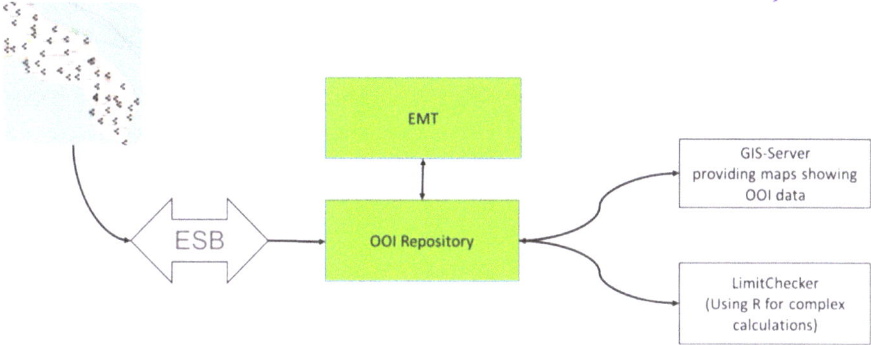

Fig. 11. EMT and OOI interfaces [3]

More precisely, the component gets already defined limit definitions from the OOI data repository, while permanently listening if there are new data available, i.e., either new limit definitions or new sensor values. New arriving sensor values will be automatically verified against the limits either by the LimitChecker itself or the verification will be forwarded to the DeployR Server [14] as shown in Fig. 12. The usage of R (as referred to in R Foundation for Statistical Computing [15]) via a connection to the DeployR Server allows even more complex calculations, modelling, verifications and data analysis, e.g., to check if there is a recognizable trend in a data series that could result in a critical situation. The result of the verification is fed back to the OOI data repository to be available not only to the EMT component, but also to all arbitrary recipients via ESB. Such architectural approach gives flexibility to the system, and availability of resources and tools for those users not using the EMT tool, but using their own proprietary tools.

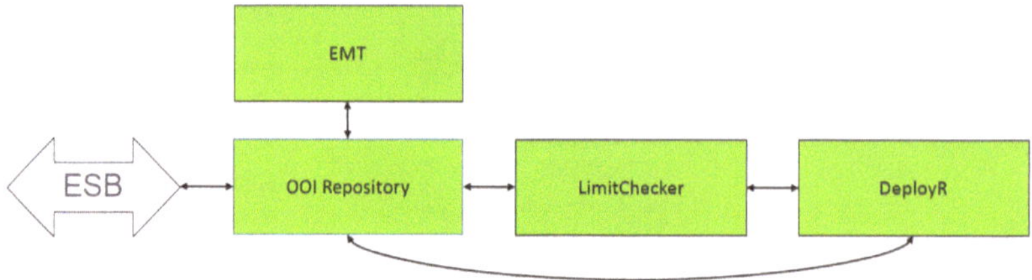

Fig. 12. Limit verification using DeployR [3]

6 Conclusions/Summary

The Emergency Maps Tool in combination with its Object Of Interest data store is an effective instrument for collaboration, management and visualisation of crises situations. It is designed to be highly flexible and customizable, thus configuration and adaptation to the user's needs can be quickly applied.

As mentioned above the design of EMT is highly flexible and modular, embedded in the knowledge layer of the C2-SENSE architecture it allows the presentation of all resources to be managed during an emergency or crises event using one a single user interface. Analytical functionalities of the LimitChecker relieve users from performing time and resource consuming activities, like monitoring of critical sensor values during a flooding event and alarm responsible persons and organisation whenever needed.

Acknowledgments. The research leading to this paper has been performed in the scope of C2-SENSE project. C2-SENSE has received funding from the European Community's Seventh Framework Programme (FP7/2007-2013) under grant agreement number 607729.

References

1. Mayer-Schönberger, V.: Emergency Communications: The Quest for Interoperability in the United States and Europe. KSG Faculty Research Working Papers Series RWP02-024; John F. Kennedy School of Government, Harvard University, March 2002. https://www.hks.harvard.edu/publications/emergency-communications-quest-interoperability-united-states-and-europe
2. Ansell, C., Boin, A., Keller, A.: Managing transboundary crises: identifying the building blocks of an effective response system: managing transboundary crises. J. Conting. Crisis Manag. **18**(4), 195–207 (2010)
3. Schimak, G., Kutschera, P., Duro, R., Kutschera, K.: EMERGENCY MAPS TOOL – facilitating collaboration and decision making during emergency & crises situations. In: Sauvage, S., Sánchez-Pérez, J.-M., Rizzoli, A. (eds.) Proceedings of the International Environmental Modelling and Software Society (iEMSs), 8th International Congress on Environmental Modelling and Software, Toulouse, France (2016). http://scholarsarchive.byu.edu/iemssconference/2016/Stream-D/106
4. Wattegama, C.: ICT for disaster management. United Nations Development Programme-AsiaPacific Development Information Programme (2007)

5. Duro, R., Schimak, G., Bojan Božić, B.: C2-SENSE: the emergency interoperability framework and knowledge management. In: Geospatial World Forum INSPIRE Conference, INSPIRE 2015, Lisbon, Portugal (2015). http://c2-sense.eu/wp-content/uploads/2014/07/Refiz-Duro.pdf
6. Gençtürk, M., Arisi, R., Toscano, L., Kabak, Y., Di Ciano, M., Palmitessa, A.: Profiling approach for the interoperability of Command & Control systems with sensing systems in emergency management. In: Proceedings of the 6th Workshop on Enterprise Interoperability, Nîmes, France (2015)
7. Gencturk, M., Duro, R., Kabak, Y., Božic, B., Kahveci, K., Yilmaz, B.: Interoperability profiles for disaster management and maritime surveillance. In: eChallenges e-2015 Conference, pp. 1–9. IEEE (2015)
8. United Nations and Office for the Coordination of Humanitarian Affairs (UNOCHA): Humanitarianism in the network age including world humanitarian data and trends, United Nations Publications (2013)
9. Richardson, L., Amundsen, M.: RESTful web APIs. O'Reilly Media, Sebastopol (2013). ISBN 978-1-449-35806-8. Accessed 29 Mar 2016
10. Apache Kafka: Open source high performance publish-subscribe messaging system (2016). http://kafka.apache.org/. Accessed 29 Mar 2016
11. OGC WMS: OGC Web Map Server. Standard interface offering map layers. Actual WMS standard version 1.3 is the same as ISO 19128 (2016). http://www.opengeospatial.org/standards/wms. Accessed 29 Mar 2016
12. OGC WFS: Web Feature Server, standard for an interface offering map layers. This layers are delivered as vector data, usually available in a lot of formats (2016). http://www.opengeospatial.org/standards/wfs. Accessed 29 Mar 2016
13. GeoServer: The open source reference implementation of a lot of the OGC standards including WFS and WMS (2016). http://geoserver.org. Accessed 29 Mar 2016
14. DeployR is an integration technology for deploying R analytics inside web, desktop, mobile, and dashboard applications as well as backend systems. https://deployr.revolutionanalytics.com. Accessed 03 Apr 2016
15. R Foundation for Statistical Computing, R.C.: R: a language and environment for statistical computing, Vienna (2015)

Visualization River Water Level Using Internet Technologies

Rafał Kłoda[1(✉)], Jan Piwiński[1], Roman Szewczyk[1], Anna Ostaszewska-Liżewska[2], and Kamil Duchna[2]

[1] Industrial Research Institute for Automation and Measurements, Al. Jerozolimskie 202, 02-486 Warsaw, Poland
rkloda@piap.pl

[2] Institute of Metrology and Biomedical Engineering, Warsaw University of Technology, sw. A. Boboli 8, 02-525 Warsaw, Poland

Abstract. In this paper we present the web-service development for water level visualization in WebGL technology, which was one of the solution proposed to be used for flood management in C2-SENSE project. The paper investigates the various technologies for visualization in different environments (groundwater, surface water, different water flow regimes and climatic zones) and their implementations in real application. Furthermore we focus on particular example taking into account whole cycle of engineering project, starting from domain analysis, survey on similar solutions and applications, project development, solution evaluation, conclusions and chart the further development plan.

Keywords: WebGL · Web services

1 Introduction

The aim of the study is to develop a system design for the visualization of river level measurement results and its implementation using Internet technologies. This system was intended to be used to visualize the results of measurements of the Fortore River in southeastern Italy in the Apulian region (in Puglia) during the first integration phase of the C2-SENSE project. This river is characterized by a fairly large drop, which in case of precipitation turns into a fierce river, causing frequent floods. The C2-SENSE research project includes the development of a reliable communications platform that utilizes current standards and network services to expose command and control (C2) functionality, sensors systems and other early warning systems, thus ensuring their interoperability for disaster management.

One of the goal of this project is to design the pilot application, to verify the actual usefulness in the field of pluviometric emergency management. For this reason it has been defined a pilot scenario that will be implemented in the territory of the Puglia Region in Italy.

R. Szewczyk and D. Havlik (eds.), *Recent Trends in Control and Sensor Systems in Emergency Management*, Advances in Intelligent Systems and Computing 675,
https://doi.org/10.1007/978-3-319-70452-4_2

The C2-SENSE was built to enable the exchange of data using prominent wired and wireless communication mediums in the emergency domain as well as will provide interoperability with the several different wireless transmission technologies and enhance the connectivity between various networks and devices.

The basic scenario during the pilot will be based on following of the Integration action of different radio networks (e.g. GSM, ZigBee) and acquisition the data from these networks to one local server to send these data directly to C2-Sense system. In this case, SQL server in the central site in Bari, have the connection with C2-Sense system, it is aware about our aforesaid data [1].

As part of the C2-SENSE project, all possible extensions and universal solutions have been foreseen. For this reason, Internet technologies such as HTML5, WebGL, Web Services and necessary adapters have been selected to facilitate later integration.

Figure 1 shows an overview of the early warning system operation.

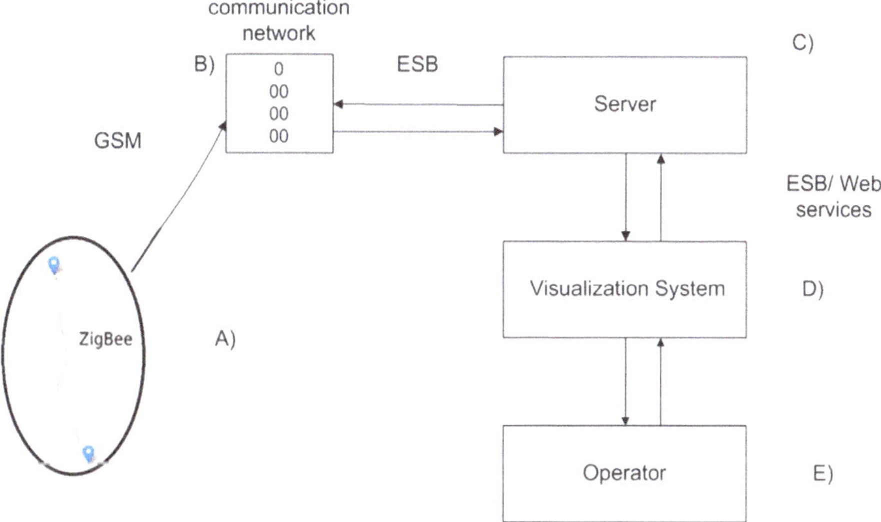

Fig. 1. The early warning system diagram: (a) river level measurement network, (b) communication network, (c) server, (d) visualization system, (e) operator

The basis of the system is a network of sensors that control the level of a river and communicate with each other and the server (Fig. 1a). ZigBee interface was used for communication between measurement modules. It specifies the wireless communication protocols described in [2, 3], therefore interface is characterized by the following features:

- Range - on average up to 100 m (maximum up to 500 m).
- Sufficient battery power is 1 year.
- Simplified transmission protocol.
- Low bit rate (depending on the bandwidth: 20 kbps to 250 kbps).

- Low cost of manufacturing or attaching a communication module to manufactured equipment.
- Maximum number of devices - over 65000.

It is used to create a mesh topology, i.e. where each node is connected to each other network node (full mesh) or different but still large number of internal network connections (partial mesh). This will ensure high reliability and will reduce the impact of single device each node will contain a module for GSM communication with the server. However, it is anticipated that prior to communication with the server, there will be internal network communication to collect results from the entire network. The result report will then be sent to the server from one module. This will reduce power consumption, extend module life and reduce operating costs while maintaining the highest possible reliability.

The server is responsible for archiving and analysing the results and predicting possible hazards and warning in situations of health and life threat (Fig. 1c). It is also planned to develop an expert system with a decision-making system for operators. The server also serves as a central place to coordinate all actions taken to eliminate or reduce the damaging effects of river activity, such as flooding and local flooding.

Due to the fact that man is assimilating and analysing most visual stimuli, an attempt has been made to develop a coherent interface for visualizing the results and positioning of the entire sensor network. Thanks to the use of Internet technology, after the necessary security and authorization, it will be able to display the measurement results anywhere with Internet access. In order to display and retrieve collected current and past measurement results, the visualization system will communicate through the Internet services (Fig. 1d, e)

2 Project Assumptions

2.1 System Technical Requirements

Due to the fact that the application will operate under a larger project, a certain framework for communication between the various elements of the system has been imposed. The exchange of data from the archiving server with the visualization server is done through web services using HTTP and XML technology. There are no imposed requirements that take into account the hardware and operating systems used, so their existence will only be a derivative and result of the chosen technology implementation. The project assumptions are presented below, broken into client and server. The client component is a visualization system that is designed to be easy-to-use and easily available on most modern electronic devices, such as PCs, tablets, smartphones, etc. In this situation, one of two solutions can be used to share data with the device. The first is the creation of a single server with a fixed communication protocol, but with different end-user implementations depending on the platform. This allows for the best matching to the capabilities and ergonomics of the device, but requires the creation of separate programs for each architecture. The second approach is to use web services, so that it is possible to get as close to the functionality and only create two implementation (desktop and mobile version). The whole interpretation is controlled by the web browser. Due to standardization most of the behaviour is identical to popular browsers such as Firefox, Chrome, Opera, Safari, Internet Explorer/Windows Edge. Because of greater versatility, the second approach has been chosen, i.e. HTTP and WWW based communication protocols and with browser support. It has also been found that the most innovative will be achieved using HTML5 and WebGL technology. Other possibilities were Flash technology and server-side image generation. Due to low mobility, Flash has been discarded. It is not supported on Apple mobile devices (iPhone, iPad) and Android. On the other hand, generating images and transmitting them from the server side creates many server-side performance issues (when handling multiple requests over a small period of time) and the speed of transmission between the client and the server.

The server part is responsible for communicating with the main server storing and analysing data. Information will be retrieved through Web Services, and the server of the visualization system will send requests for data. It was decided to write a server component in Java using Tomcat as an application container. The additional software used will be the HTTP server - Apache. This kit was chosen for the sake of Java's advantages:

- Portability - no application server dependency on the installed operating system.
- Ease and speed of writing code - There are many integrated environments for writing code (e.g. Eclipse, NetBeans, IntelliJ).
- Popularity - Java has become the dominant programming language in recent years. An additional factor influencing the choice of language was the professional practice of the author of the work in the aforementioned technologies.

The input data of the program will be data downloaded from the server. The basic input is a tuple: sensor name (identifier), location (latitude and longitude), measurement result (depth), and date of measurement. The output of the program will be a map with a detailed river bed along with marked measurement results. In conclusion, the following technologies and solutions have been proposed:

- Client side support via browser.
- Internet technologies: HTML5, WebGL.
- Implementing server parts in Java using Apache and Tomcat.
- Communication between servers via Web Services.

3 Implementation of an Application

3.1 Selected Technologies

The main premise for the choice of technology for systemic implantation was to base it on ready-made, technologically stable solutions. This applies to both the selected programming language and the integrated development environment (IDE). Below is a description of the tools used to perform the job:

- Programming Language: Java - is an object-oriented programming language. Source programs are compiled into bytecode, that is, the form performed by the virtual machine. It fulfils several paradigms: object-oriented, structural, and imperative. It was chosen for its portability, its author's experience in it, great integration with web services, and a number of documentation supporting work. Java is available under the GNU General Public License.

- Developer Platform: Vaadin - is a Java application framework that offers a range of Rich Internet Application (RIA) applications. A dynamic one-screen interface is envisioned. Using Vaadin eliminates the hassle of standard HTML technology. It offers server-side architecture and automatically supports AJAX technology. The code is written once, and at compilation I am separated into server and client (Widgetsets). Free software license: Apache License 2.0.
- Mapping library: LeafletJS - was selected as the most transparent, with an intuitive map library support. In addition, it has provided support in the form of an extension in the Vaadin platform via the official add-on. It uses HTML5 and CSS5 technology. The first edition took place in 2011. Free Software License: BSD-2-Clause.
- 3D visualization library: Cesium JS - one of two found libraries supporting WebGL technology. No official stable addition to the Vaadin platform. Library support will be provided through the JavaScipt API. License to Apache License 2.0.
- Application container: Apache Tomcat - has been selected as an application/server container to run Web applications in Java Servlets and Java Server Pages (JSP) technologies. It is available under a license: Apache License 2.0.
- IDE: Eclipse - Eclipse was written for writing applications. It provides a comfortable and stable development environment. In addition, it has integration with Maven - a tool that automates compiling and building software on the Java platform. License to Apache License 2.0.

The developed system will use directly or indirectly through libraries the following programming languages and technologies: HTML5, JavaScript, CSS3, SCSS, Java, AJAX, Web Sockets. In addition, if the program was made available on the Internet, the HTTP server should also be used. The proposed option is Apache HTTP Server. Available under license: Apache License 2.0. The selection of the development environment included both the convenience of software development, its stability, development opportunities and licensing issues.

3.2 Application Description

The application was implemented using the technologies indicated in Sect. 3.1.

A dynamic, one-screen graphical user interface (GUI) has been implemented. Intuitive map control and a clear menu are used. In relation to the original concept, the main menu was moved from the right to the left. This greatly increased the ergonomics of working with the program. The top menu was also dropped. Options moved to one menu on the right. This expanded the map area and raised aesthetic qualities. Figure 2 shows a cutting from the main program window.

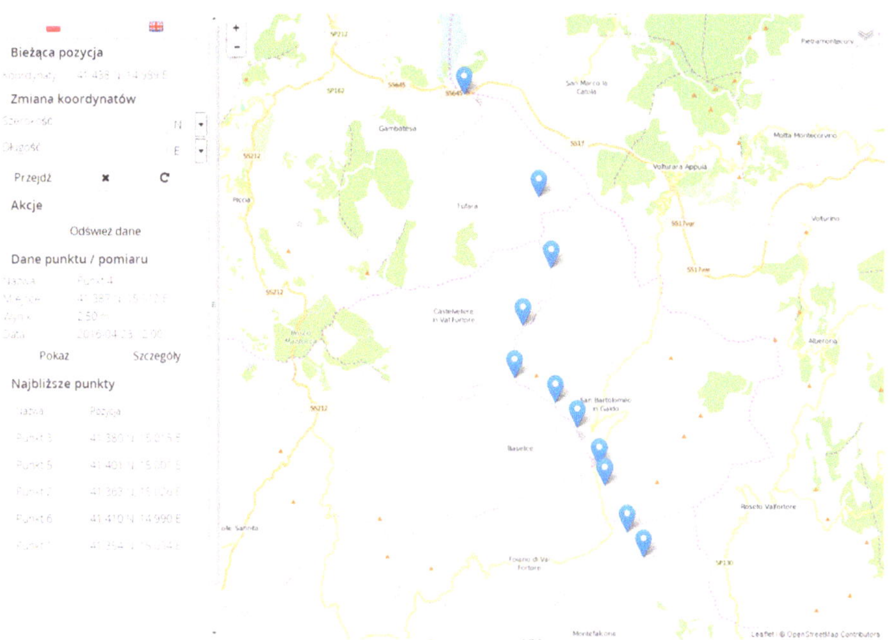

Fig. 2. Cutting from the main window of the implemented application

The program occupies the entire available surface of the page both in width and height. In the central part there is a map. The left side occupies the menu. The breadth of menu widths to maps is 1 to 3. This provides the ergonomics possible with the map and provides clear measurement information. Information coupling and actions performed on the map menu are guaranteed. For example, clicking on the selected element of the list of neighbours moves the map view to a given point. The following functionalities are displayed in the menu:

- Change of application language (Polish and English).
- Current position.
- Change of coordinates.
- Actions (refresh data).
- Measurement point data.
- List of neighbours.

The current position displays the coordinates of the center of the map window. GPS data displayed are rounded to 3 decimal places, therefore this allows to find ones position in case of too quick movement on the map. Changing coordinates is used to change the position on the map if coordinates are known. They are set separately for longitude and latitude. The default settings are aligned to the C2-SENSE project, that is, the location of the Fortore River. After entering the desired values, click the "Go" button. There are also 2 other buttons with "X" and a rounded arrow icon. The first action cleans to force the program to download the latest data and measurement data displays current results and information for a given location. It is possible to change a point by clicking on the selected marker on the map or by selecting a point from the neighbourhood list. Displayed data is: name, location and last measurement data: score and performance. There are also two other actions available: show and details. The first is to go on the map to the point. It does not change the zoom level, only the camera setting. The second action enable to open the details on the window, as is showed in the Fig. 3, below. The nearest panel shows list of five closest neighbours, as showed in Fig. 2 in left corner. These distances are measured in a vector manner between the points. This list allows to quickly move to the next measurement points, by choosing a neighbour by right-clicking on a row in the table and it causes the map to move to the selected point and to display its basic data in the measurement panel.

Fig. 3. Detail window of application

The map contains a zoom panel in the top left corner. The mouse wheel may be used to change the zoom level. In the top right corner there is a possibility to change the maps provided. Currently, maps are provided by OpenStreetMap (OSM), Esri WorldImagery, Mtb Map, and OpenTopo. However, there is an easy possibility to introduce others according to ones needs. In the lower right corner there is information on the use of the Leaflet and the map provider in someone's project. On the map there are markers indicating the position of the measuring point. Operator can click on them, to move the map position so that the point is in the middle. In addition, a balloon with point information (name, result and place) and the ability to open the details window are displayed.

The details window is invoked by clicking on the details button on the point data panel or by running from the balloon located at the points on the map. The window opens in the center and in modal mode. This mode greets data outside this window and prevents interaction with them other than by closing the window. Once closed, the application returns to the base state where data and maps are available. The details window is divided into 2 parts: point data and analysis. The presented data includes:

- Name.
- Location (GPS coordinates).
- The result of the last measurement.
- Date of last measurement.
- Value of the warning level.
- Alarm level value.
- Nearest points.

The list of nearest points allows to quickly move to neighbouring measurement points. The analysis section contains tabs with graphs and visualizations made using WebGL technology. The range of presented measurement dates and their results is determined by the date filters. Dates cannot cross, i.e. the start date cannot be longer than the end date. In the event of an intersection, the system automatically sets the dates to equal. Below the chart are presented possible actions to perform. These are refreshing and exporting. Refresh analogy to actions on the map, retrieving the latest data from the server. Export generates a PDF file containing the displayed point and graph data. Charts are executed in the JfreeChart library, export to PDF is done using the iText library, and visualization in the Cesium JS library.

4 Conclusions

As part of the implementation work, the core and backbone of the system were developed. It is a good base for developing it depending on further requirements and needs. Here are briefly implemented functionalities:

- Layers: background (map) and measurement results.
- Intuitive ability to move around the map.
- Clear user interface.
- Ability to quickly move to measurement point (menu list).
- Import measurement results.
- Generating graphs of measurement results due to user-specified time interval.
- Multilingualism (Polish and English).
- Basic 3D visualization using the Cesium JS library in the tab detail.

Thanks to technology used in our developed application, navigating the map is intuitive and ergonomic. Furthermore, it allows Operator to add more visual layers, for example the current location of the fire department. Common idea about described application, in this paper, is based on open architecture, which consists universal adapters, structural design pattern, therefore it enables the connection to any sensor data source. Overall, this solution has the primary goal to design a cooperation platform that promotes spatial data interchange involving the end user, and provides support for the assessment and decision on the territory to determine how to deal with in operating mode and systematic different types of events from the knowledge of the location, by the compliance Geographic, from environmental data and historical information.

This goal represents an improvement from a basic approach focused on damage management, to a culture of prevention and prediction, spread out at various levels, based on the identification of risk conditions and the adoption of measures aimed at minimizing the impact of events.

References

1. Bączyk, A., Piwiński, J., Kłoda, R., Grygoruk, M.: Survey on river water level measuring technologies: case study for flood management purposes of the C2-SENSE project. Advances in Intelligent Systems and Computing, vol. 543, pp. 610–623, Springer (2017)
2. Franceschini, F., Galetto, M., Maisano, D., Mastrogiacomo, L., Pralio, B.: Distributed Large-Scale Dimensional Metrology. Springer, London (2011)
3. Nawrocki, W.: Measurement Systems and Sensors. Artech House, Norwood (2005)

Profiling Approach for the Interoperability of Command and Control Systems in Emergency Management: Pilot Scenario and Application

Marco Di Ciano(✉), Domenico Morgese, and Agostino Palmitessa

InnovaPuglia, Valenzano, Italy
m.diciano@innova.puglia.it

Abstract. In order to manage emergencies, crises and disasters effectively, different organizations with their Command & Control (C2) and Sensing Systems have to cooperate and constantly exchange and share data and information. In other words, territorial emergency management requires a cross-organisational, cross-domain, cross-level interoperability between the involved C2 and Sensing Systems. Although individual standards and specifications are usually adopted in C2 and Sensing Systems separately, there is no common, unified interoperability specification to be adopted in an emergency situation, which creates a crucial interoperability challenge for all the involved organisations. To address this challenge, we introduce a novel and practical profiling approach, which aims at achieving seamless interoperability of C2 and Sensing Systems in emergency management. Unlike the conventional profiling approach, which addresses only first three layers of interoperability stack, the profiling approach introduced in this paper involves all the layers of the communication stack in the security field. The work presented in this paper examine in particular the aspects relating to the testing of the project in the region of Puglia and interfacing with information systems of local authorities.

1 Introduction

C2-SENSE is a system that allows alignment and cooperation between all entities involved in emergency management in the event of natural disasters such as floods. Effective management of emergencies, crises and disasters depend on information readily available, reliable and interchangeable. To achieve this, many different organizations with different "Command and Control (C2) Systems and Sensors" must cooperate through interoperability [1].

The concept profile is intended to eliminate the need for a bilateral agreement between any two partners to exchange information by defining a standard set of messages/documents, choreography, business rules and constraints. The partner profile

InnovaPuglia Company subjected to the management and control of the Puglia Region.

R. Szewczyk and D. Havlik (eds.), *Recent Trends in Control and Sensor Systems in Emergency Management*, Advances in Intelligent Systems and Computing 675,
https://doi.org/10.1007/978-3-319-70452-4_3

compliant are able to exchange information and services between them. Considering the nature of emergency management, where organizations can change at runtime must allow flexible coordination to address unforeseen circumstances and to prevent the chaotic response to crisis situations.

For emergency management, organizational aspects, such as policies, procedures, operations and the strategies are as important as the technical aspects of interoperability.

C2-SENSE is an ambitious project, on the basis of procedures open source software and existing standards. This will facilitate the development efforts and help you easily identify gaps where there are new technological solutions, guidelines, recommendations or standards. In the scope C2-SENSE builds a platform of cooperation that promotes the exchange of spatial data involving the end user, and provides support for the evaluation and decision on the territory to determine how to deal with mode of operation and systematic different types of events to the knowledge of the position, by compliance geographic data from environmental and historical information.

2 Interaction Between Involved Stakeholders

Daily, the CFD issues a bulletin about the regional criticalities which is transmitted by email to the Regional SOIR and Operations Rooms of the neighboring regions that is Molise, Campania and Basilicata [2].

The alert message is forwarded via fax through a multi-channel platform to:

- Prefectures
- Provinces
- Municipalities
- ASL (Health Department)
- DPC (National Civil Protection Department)
- Air Force
- Regional Directorate VVF (Fire Department)

  which are present in all the areas affected by the critical alert.

Through the Press Office of the Regional Council, all the main regional newspapers are reached. The Prefecture office represent the government in the province territory and activate the hospitals and military bodies. The network of volunteer organizations is activated by the Regional Operations Room or by the COC (Municipality Operating Centre).

3 Criticities

C2-SENSE project aims to use ICTs to improve collaboration between the various crisis management functional groups, local authorities and emergency workers, making the sharing of digital data possible among their standalone systems through the profiles of interoperability and a platform where participants, which conform to these standards-based profiles, collaborate effectively. In the development of these profiles for interoperability, C2-SENSE will address the following emergency current:

- Protocols and standardized procedures
- No good mechanisms for merging the related information
- Lack of coordinated concept of operations
- Limited awareness common situation
- Inadequate support procedural for crisis operations joint
- Inability to partners' attention to relevant information because of its lack of persistence
- Automation Support
- Awareness Team
- Inadequate capabilities for sharing classified and sensitive information
- Information (and the tools used to convey it) is not always reliable
- Senior management facilities
- Inadequate joint training programs.

4 Pilot Scenario

The two-days Pilot Scenario describes what could happen step by step, before and during a flooding along the Fortore River as well as the probable evolution that would occur.

The Pilot scenario involves different stakeholders described in Sect. 2.

During the first day, the Pilot Scenario describes what are the institutions involved in the "Forecasting Phase" and what are the documents produced [3]. In particular:

- The National Weather Service (CFN) makes a prevision of bad weather conditions for the next 24–36 h;
- The Regional Functional Center (CFD) issues a Bulletin about the regional criticality.
- The manager of Regional Civil Protection publishes and sends an Alerting Message to Prefectures, Municipalities, etc.

As a result, other actors are involved during the first day of the scenario: Prefecture of Foggia, Province of Foggia and local Municipalities.

During the second day, the Pilot Scenario describes what could happen during a flooding along the Fortore River, what are the institutions and organizations involved and their roles and responsibilities.

In particular:

- CFD (Regional Functional Center) follows the evolution of the situation through the regional monitoring network in the territory (monitoring and surveillance activities) [4].

- SOIR (Regional Operations Room) keeps in contact with Municipalities and allows land monitoring through municipal structures of Civil Protection as well as through Voluntary Organizations. It ensures emergency operations and H24 service.
- The Prefecture opens its Assistance Coordination Center (CCS) at a provincial level with the presence of Healthcare Services, Police Department, Fire Service, Red Cross, etc.
- Municipalities, the responsible body in their territory, activate its own emergency procedures (Civil Protection Plan). They also activate the Municipal Operational Centres (COC) by giving notice to the SOIR and to the Prefecture of Foggia. Municipalities also keep in contact with the Voluntary Organizations [5].

5 Pilot Application

In Fig. 1 Architectural design, you can see the architectural scheme that summarizes the key features of the project C2-SENSE, and correlates them with the systems of the organizations/local authorities/stakeholders in emergency management situation. The diagram considers the technology currently present in the systems of the involved actors, while giving designers the opportunity to replace them with other technologies available on the market.

Therefore some of the systems shown in the following figure will be simulated either for technical reasons or for the needs of the involved institution/organization.

Functions related to C2-SENSE can be grouped into three main categories:

- Integration profile
- Action Management
- Communication routing.

Integration profile is the capability to configure C2-SENSE using customized profiles to describe emergency procedures according current regulations in the country. Action management is the capability to manage organizations' actions, and automate the processes, however, giving the user the ability to retain decision-making power. At the end, communication routing is the capability to manage and check communications between involved stakeholders.

The involved stakeholders can be grouped into three categories: the control base is made up of the civil protection systems which are considered the main user of the C2-SENSE system. For field points we can find actuators such as alarms or automatic barrier systems, voluntary organizations and citizens. Command stations, instead, are command and control points such as prefectures and municipalities, but also other decision-makers such as the fire department and the medical centers involved in the territory where the emergency occurs.

C2-SENSE Pilot Application aims to provide improvements following the introduction of the following specifications:

- New types of interoperable sensors
- Machine processable XML Alert Messages over e.g. Web Services
- Social media connection

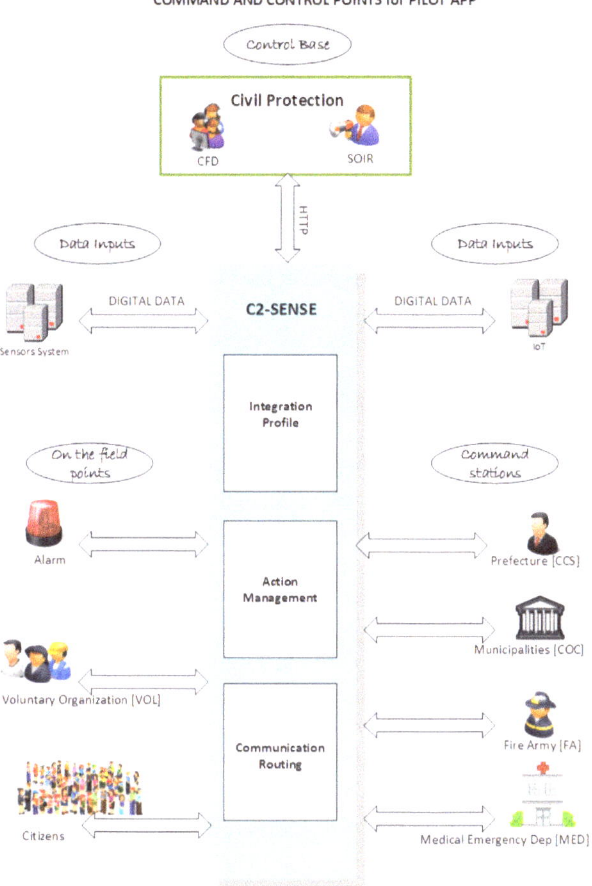

Fig. 1. Architectural design

- Communication through SMS
- Resilience and consistency of the communication channel
- DSS (Decision Support System) support.

The sensor data will be made available to the system depends strongly C2SENSE transport infrastructure, which must be available, fast and reliable. The communication infrastructure used by the service of the Puglia region of civil protection consists of a network that uses radio frequencies licensed to carry data from sensors to the headquarters of the Civil Protection of Bari.

From a procedural standpoint, C2SENSE Pilot Application brings significant improvements in the interaction of the various organizations involved in emergency management.

In particular C2SENSE allows profiling of generic integration that map all the possible solutions proposed by the system for emergency management. Due to additional phases in which the profiles are to be made, they need to:

– Allow the system to adapt to all circumstances in an emergency, from the time that the emergency situations are often unpredictable;
– Adapting to the different structure of organizations
– Change to the different functions of different organizations in different countries
– Change the different legislation in force in the country in which is managing the emergency.

After the generic profiles have been customized, they are ready to be used by the system to generate the operations that automate the procedures of communication and information exchange. In this way, the system provides a valuable guide to the following procedures according to the occurrence of a specific event of an emergency. In fact the profiles specialized for Puglia region will be executed through a specific tool. It means that organizations taking part in the emergency plan of Puglia region will exchange information among themselves according to the specifications in the profiles and C2-SENSE system will control and track these operations and display the progress.

Other improvements that have been introduced are: the addition of useful new features to improve the process of assessing the situation hydrometric, and the complex process of managing relief.

In particular, the system allows for 'real-time' input of new sensors, whose values are communicated in real time to the control center, and interpolated with the data already present.

To improve both phases of analysis of the situation and rescue, a useful help can be given directly by citizens, through the use of a special application with which they can receive information about the state of emergency directly by the organizations responsible, and citizens can give them their location or other information.

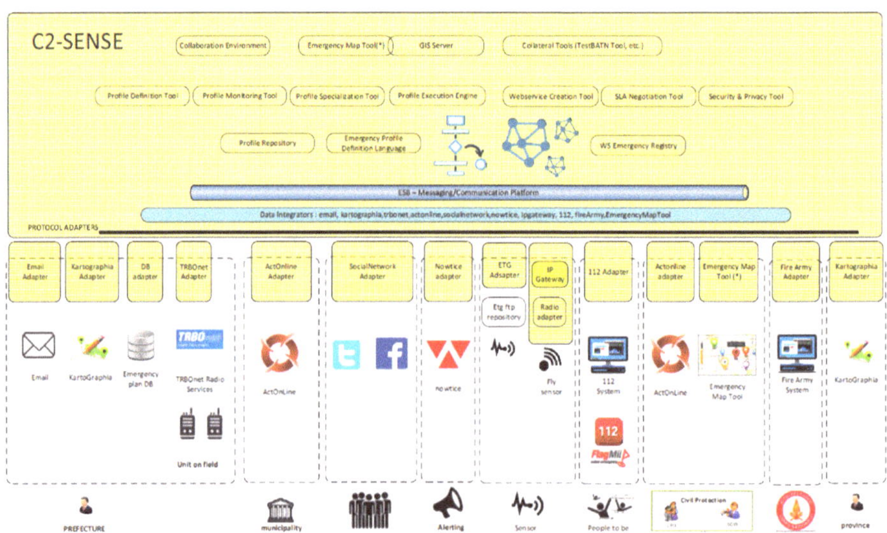

Fig. 2. Pilot application environment

In Fig. 2 the pilot environment is shown. It is divided into two different logical ecosystems: C2-SENSE test environment (on the top) and end-users systems (second half).

C2-SENSE test environment is a digital ecosystem composed of various C2-SENSE tools having specific tasks: tools for definition and specialization of integration profiles, tools for communications management between the different C2-SENSE systems, and tools for performance monitoring. This ecosystem interacts with another ecosystem consisting of all systems of the local end users involved in emergency management and taking part in the pilot application testing. The interaction between the two ecosystems is done via different interfacing modules, called adapters, implemented between end-user's system and the C2-SENSE tools. Data used by the end-user system is translated by the adapter to make it C2-SENSE compliant; then, this information is processed and transmitted to the end-user recipient according to his local understanding or standards. The integration profiles define the way in which such integration has to be done, furthermore, from a procedural point of view, the system improves the way to allow communication between the organizations involved. It allows you to send feedback to confirm receipt of a message, ensure the success of a communication.

At the end of a test phase, a validation checks ensures that the product design satisfies or fits the intended use (high-level checking), i.e., the software meets the user requirements. In other words, software verification is ensuring that the product has been built according to the requirements and design specifications.

6 Conclusion

In this paper applicative aspects related to the project C2-SENSE, with the aim of providing valuable and assessable instruments with regards to the assistance in emergency management, interoperability among the information systems of the involved stakeholders. Within the project applicative aspects have been designed using a novel profiling approach, which addresses all the layers of the communication stack in security field. In order to ensure that the developed profiles are generic and applicable in real life setting, they are being assessed in a realistic flood scenario in Puglia region of Italy. The current situation called Pilot Scenario has already been analyzed; and stakeholders are also identified. Next step will be on the filed interfacing with the real information systems and the interaction/emulation of production data, in order to create a testing environment of the whole process of emergency management. At the end of this phase we will proceed with the evaluation of results and achieved performance.

Acknowledgements. The work presented in this paper is achieved in the scope of C2-SENSE project supported by the European Community's Seventh Framework Programme (FP7/2007-2013) under grant agreement number 607729 and partly by the APPS (9140033) project supported by TÜBITAK in the scope of ITEA3 Programme.

References

1. DPCM (Directive of the Presidence of the Council of Ministers) of 3rd December 2008: Operational guidelines for the management of emergencies
2. National Law no. 100 of 12th July 2012: Urgent provisions for the reorganization of civil protection
3. DPCM (Directive of the Presidence of the Council of Ministers) of 27th February 2004: Operational guidelines for the organizational and functional management of the national and regional warning system for hydrological and hydraulic risk for the purpose of civil protection
4. DGR (Deliberation of the Regional Council) no. 2181 of 26th November 2013: Declaration of Activity of Regional Functional Centre CFD
5. Ordinance of the President of Council of Ministers no. 3606 of 28th August 2007: Operating Instructions for the preparation of a plan of inter-municipal civil protection

Partial Automatization of Legacy Systems Integration Using Web Application Creator

Romuald Périnelle(✉), François Gendry, and Christophe Guettier

Safran Electronics & Defense, 100 Avenue de Paris, 91300 Massy, France
{romuald.perinelle, francois.gendry, christophe.guettier}@safrangroup.com

Abstract. Interoperability is as old as networked applications. New contexts and technics offer new challenges and new opportunities to cope with them. This paper presents a way to use modern technics and C2-SENSE concepts of Profile to partially automatize the integration of existing systems into a standardized framework used for interconnection of emergency applications. This approach has been proven as efficient in C2-SENSE pilot application.

Keywords: Interoperability · Legacy systems · Protocol adaptation · Web service generation

1 Introduction

Interoperability problematics are a concern since the emergence of networks, and even before. Among the first domains of application was the interoperability of systems within an administration or a company: human resource management and financial applications, manufacturing and supply chain, etc. These issues never found their Grail, but the computer scientists found reasonable answers to physically interconnect computers (physical interoperability), to exchange data (protocol interoperability), and to ensure that the data was understood (syntactic and semantic interoperability). They used network technics, as Ethernet and IP, and shared data (common database, shared file systems, etc.). This first step could not survive to the increase in complexity, and was problematic as on the shelf software relied then on internal database that could not, and should not, be hacked to use a common database. Direct communications were also used, and this track led to the definition of mediators [1] and to the definition of I3 Architecture where I3 stands for *Intelligent Integration of Information* [2]. These mediators, which can be stacked and/or organized in networks, ensure each a part of the necessary operations for full interoperability. This is a *divide and conquer* strategy to treat the very complex problem of the interoperability. With these models the focus was put on semantic translation with the use of ontologies by semantic mediators.

Next step was to go out of the company. Companies needed to exchange data along the supply chain, between partners, and upward with catalogs. In the military field the concern is the increase in joined operations in parallel with the already important digitalization of command and control centers. Ensuring interoperability in this context

R. Szewczyk and D. Havlik (eds.), *Recent Trends in Control and Sensor Systems in Emergency Management*, Advances in Intelligent Systems and Computing 675,
https://doi.org/10.1007/978-3-319-70452-4_4

needs more than physical links, protocol, syntactic and semantic compatibilities. The procedures and, further, the goals, have to be shared and agreed. A company or an army will not usually accept to change its internal way of managing things just to be able to interact with a pair. This lead to models to be used in military or in civil domains, like in [3], defining interoperability layers, from *physical* to *agreed procedure*, then up to *common goal*.

This evolution has also impacts on the lower layers: network used between companies is now Internet, but aggression is common, which forces the use of firewall and which limits any protocol other than web based ones; for armies, OTAN issued standards including physical network, protocol and data model. From the 90 s, the military model is data centered with the idea of a distributed and/or duplicated database. Military interoperability proved to be very complex, and is in constant progression. [4, 5] are examples of recent advancement. The contestation of the data centric architecture and of JC3IEDM as too complex is another example, which leads to more message oriented and service oriented architectures, as in [5].

C2-SENSE takes advantage of these evolutions and uses a layered model of the interoperability (see Fig. 1: Interoperability Stack of the C2-SENSE Project), a message centered data model, and a service oriented architecture.

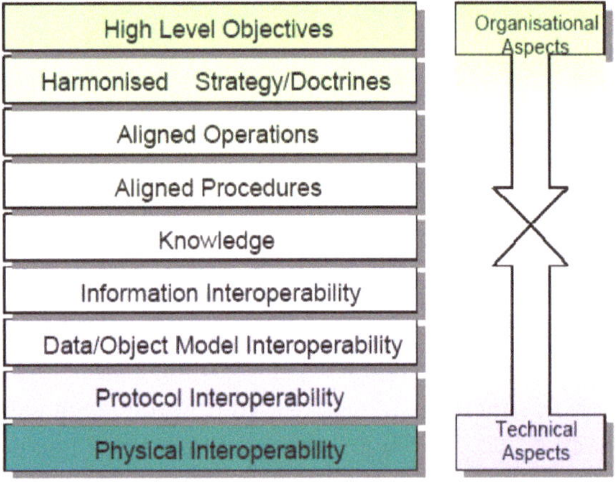

Fig. 1. Interoperability Stack of the C2-SENSE Project

While new applications can directly implement the C2-SENSE proposed standards, fact is that the involved organizations have pre-existing applications that will not easily evolve to use them. For this reason, in C2-SENSE, adaptors and mediators are used. But the protocol adaptation is both repetitive (some protocols are used by many applications, like SMTP, other protocols share common elements, like SOAP web services) and always specific (each application has its own set of messages, and uses its protocols with different configuration, from the email address to the way files are put on the FTP server). To gain productivity in this context, in C2-SENSE, we developed a semi-automatic generation of the protocol adaption. This later is here called Legacy Application Representative (LAR). The semi-automatic tool has been called Web Service Creator Tool (WSCT).

2 Architectural and Methodological Environment

2.1 Legacy Application Representative

These modules ensure the protocol, syntactic and semantic mediation for legacy systems. Protocol adaption is performed internally, while syntactic and semantic mediation is performed by calling web serviced provided by the Data Integrator (DI) and the Semantic Interoperability Suite (SIS), two specialized mediators (see below Fig. 2 LAR place in interoperability layers).

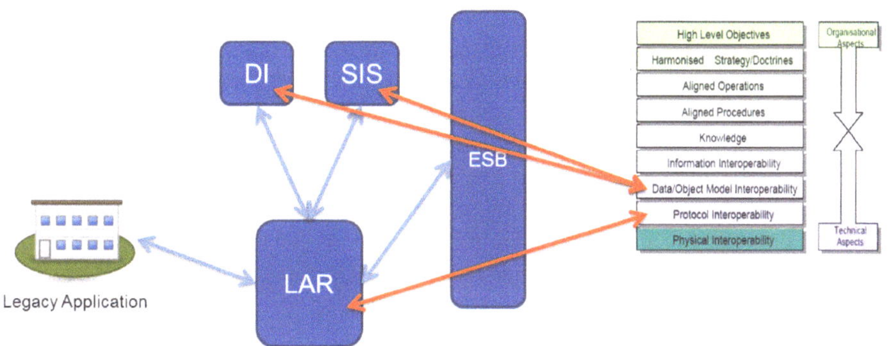

Fig. 2. LAR place in interoperability layers

Since the central way to transfer messages in C2-SENSE is an ESB in passive mode (it offers services to send and receive messages, but will never call back or notify a registered client), the LAR offers no callable web services. It sends and receives messages to and from C2-SENSE using the defined ESB.

The ESB is configured using profiles. These profiles correspond to upper layer concepts (information, knowledge, procedures, etc.). LARs will then send and receive messages to and from ESB knowing the corresponding profile.

LARs are in fact totally independent from the upper interoperability layers, and don't even treat messages content once the protocol affairs are done, which include:

- Identifying message type for message translation.
- Identifying source and destination for routing.
- Identifying corresponding upper layer profile for routing.

Each LAR manages the adaption of one legacy application, and offers to the C2-SENSE system a compliant interface representing the legacy application capabilities. Two important things have to be known about this:

- LARs are not autonomous applications, they are adaptions, they have no human interaction beside basic maintenance operations.
- LARs capabilities rely on their legacy applications. A LAR will not usually generate a message, it will request that the legacy application sends one, and then ensure its translation and forwarding. Exceptions are purely protocol messages, like transmission acknowledgment.

2.2 C2-SENSE Type Projects Phases

To well understand the semi-automatic generation of LARs, we have to well understand the main phases of an interoperability project.

When these projects start, organizations exist and may already have each their own tools and procedures. They often already communicate with each other, using more or less evolved tools, ranging from phone and fax, up to integrated solutions.

The phase 1 begins: in this phase, organizations are listed, as well as their tools, habits, procedures, and goals. For Emergency applications, C2-SENSE proposes to use predefined generic profiles as abstraction of usual procedures and exchanged information. In the phase 1, these profiles are specialized, which means that the procedures are described using the generic bricks and replacing the generic roles by precise organizations. Then, the interoperability tools like LARs, Data Integrators, Semantic Interoperability Suites, ESB, etc., are built, configured and run.

Once all is running, phase 2 can begin: organizations can send and receive messages through C2-SENSE architecture which takes care of routing, translation, and of verification of the conformity to agreed procedures.

In case of evolution of individual organizations, arrival of new ones, or renegotiation of responsibilities, the project can switch again to phase 1, or, better, a partial redesign (phase 1) can be anticipated while the applications are still running (phase 2).

The definition of generic profiles and of the applicable standard has been called Phase 0 (Fig. 3).

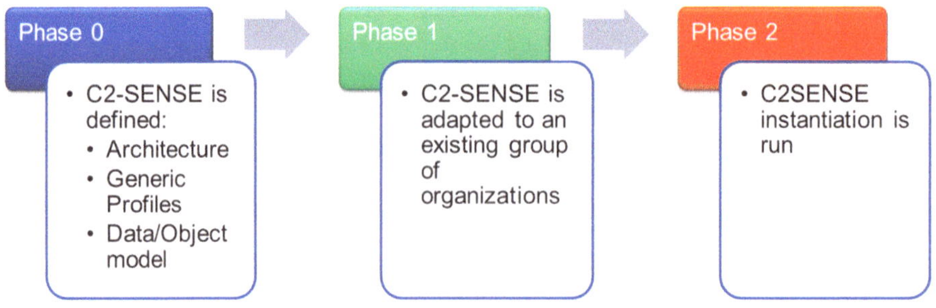

Fig. 3. C2-SENSE phases

3 Semi-automatic Generation

3.1 Building LARs

LARs are built late in phase 1 and used in phase 2.

In the following figure, the phase 1 is detailed (Fig. 4):

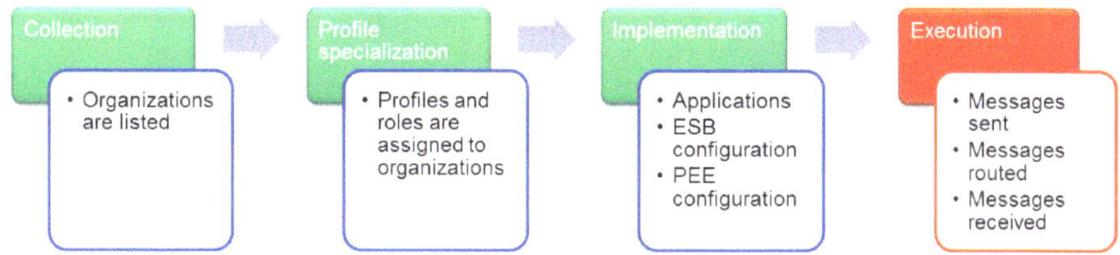

Fig. 4. Phase 1 details

LARs are part of applications: the application in the sense of C2-SENSE is there the couple Legacy application plus its Legacy Application Representative (LAR).

To build a LAR, we need:

- Generic profiles.
- Specialized profiles.
- Legacy application protocol definitions and configurations.
- ESB configuration (mapping of profiles on it, ESB URL).
- Data Integrator and Semantic Interoperability Suite URLs.

In C2-SENSE architecture, profiles, generic or specialized, are provided by the Profile Definition and Specialization Tool (PDST). Protocol definitions and configurations are provided by human operators or experts.

To help building the LARs, a common skeleton has been defined, with agreed interfaces. In this skeleton, modules have been identified, among which:

- Legacy protocol adapter.
- DI proxy.
- SIS proxy.
- ESB proxy.
- A Kernel to organize all that.

The legacy applications protocol may or may not exist in the set of already developed protocol adapters. Proprietary protocols need specific developments. Once this is done, building a LAR is a matter of configuration and of building an application from existing libraries (Fig. 5).

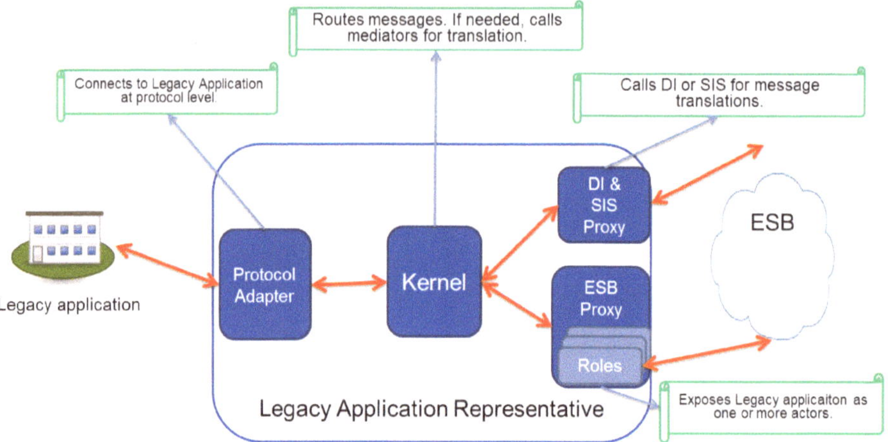

Fig. 5. LAR architecture

3.2 A Web Service to Build Web Services

The Web Service Creator Tool is the application that we developed to semi-automatically create, upload, and run LARs. This is a web application that builds web applications (Fig. 6).

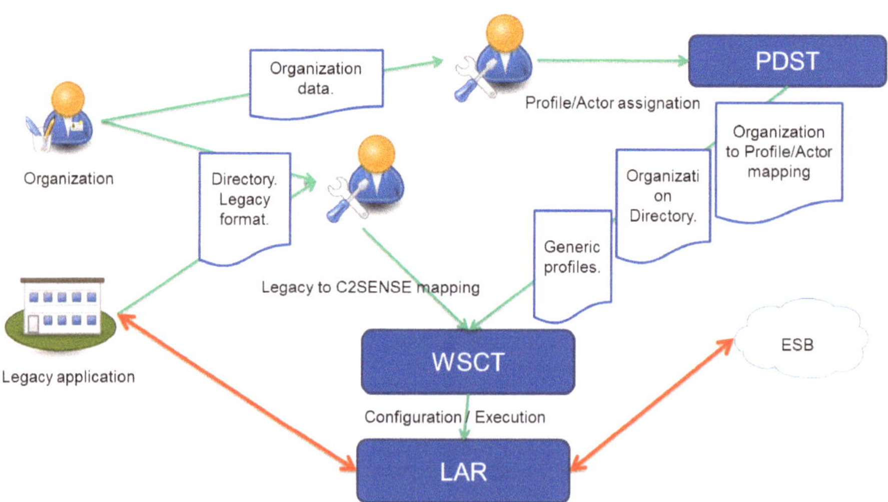

Fig. 6. WSCT creating LAR

This tool can, on request of the user, download the profiles from the PDST. From these data, it knows the list of declared organizations, and their associated profiles, as well as all generic profiles, used messages, and so on.

The WSCT embeds a catalog of protocol adapters. The current version of the tool handles emails (SMTP), Twitter (publish only), FaceBook (publish only), Act Online proprietary protocol, a proprietary radio system (TRBONET), and Atom/RSS (including proprietary specificities for a given Fire Brigade). This can easily be completed.

The user can create a LAR and give him a name. Then he can assign an organization, which will automatically select profiles, messages, etc. In parallel, he will select a protocol adapter and configure it.

Once this is done, the LAR can be uploaded on an application server. The WSCT can be configured to manage many application servers, using secured access. Then the LARs load can be balanced on various physical or virtual servers.

4 Main Overcome Difficulties

The tool is as simple to use as it could be difficult to design.

First difficulty was to design properly C2-SENSE so that necessary information was available at the right place. A good architecture gives simple solutions. The notion of profiles helped a lot to simplify the design of the WSCT, as the expected behavior of the legacy applications have been abstracted and limited to a given known set.

Protocol adaptation itself is a subject. The job has been done many times, but here we need something more generic as the protocol adapter is selected in a library from an extendable set of adapters. We had to get to the very concepts of protocols: what they are for, what services they offer, etc. For example, common services are:

- Connection and keep alive.
- Acknowledgments (reception, possibly reading).
- Warranty of transmission (repetitions, etc.).
- Thread identification (responses linked to requests).
- Multiple recipients.
- Carbon copy.

The central service is always to transmit higher level messages and get answers.

The LAR kernel had to manage most of these elements in a protocol agnostic way. *Warranty of transmission* is not the concern of the kernel. *Multiple recipients* is to be handled because it is linked to message routing. Our works demonstrated the feasibility of such kernels.

Many legacy applications are very limited compared to ideal profile definitions. This is a real issue for upper layer interoperability as the necessary data cannot be extracted from their messages. This is not an issue for the WSCT and the LAR as long as the message type can be identified (this is the job of upper layer mediators), but even message identification is a real issue.

In fact, no software application can invent missing information. The use of default value to fill an incomplete message is possible (and we used that trick), but may lead to misunderstanding as the destination can interpret the default value as intentional. As the software community says: *garbage in, garbage out*. As rough conclusion, already known for a long time in military interoperability systems: *to make systems interoperable for a given set of features, you need systems that implement these features.* For an emergency interoperability project, the first step is to ensure that every organization owns a tool that offers the basic features that you'll have to make interoperable. You need that the Police can request an ambulance? The police organization needs a tool that let him request for an ambulance, and this application have to be able to transmit this request in a way or another. Once this is done, the interoperability project will be able to ensure that the message is got, translated, transmitted, and translated again up to the Ambulance system – which, obviously has to own such a system which can accept requests. This is not an issue to be overcome by a LAR. This is an issue that has to be overcome by the whole interconnection and interoperability project.

The used architecture implies that the LAR can recognize the message type when it receives a message from a legacy system. Since the list of possible message types is not known by applications that existed long before C2-SENSE abstraction, the general case is that either the list of message types differs, or there is no message type at all in the legacy application. To solve this, the following strategies are applicable:

1. modify legacy system: this is to be avoided in many cases;
2. or change the organization's procedure so that messages sent by the legacy system can be recognized. For example, an unused field will be used to set the message type, or a specific string will be added in the subject, and so on;
3. or use complex rules and a semantic mediator to map the incoming message to standardized types.

In our project we focused on solution 2: when message types could be recognized, simple rules were used, otherwise the use of the legacy application was slightly changed so that the common procedures could run.

For example, an organization used simple mail. This application can generate messages where answers could be recognized, as well as acknowledgments, but no high level message type can be recognized. As the common procedures imply that this organization can emit a Mission Plan, and as the mail is plain text, the organization procedure had to change so that mission plans were written with necessary information, including the information that this was a mission plan.

Other example, the Fire Brigade uses a more elaborate tool, which emits and receives CAP messages over RSS feeds. These messages are formatted and fields can give indication on the message type. Unfortunately, CAP format is very specialized in alert transmissions, while the Fire brigade needed to receive and send other types of messages. The message type and some specific data had to be put in unused fields and/or in message comments.

5 Conclusions

We demonstrated the feasibility of a web service to generate Legacy Application Representatives (LARs), that is to say adapters that will connect legacy applications to a common framework and expose their services as if they were compliant to the framework standards.

This application generator strongly uses modularity and protocol abstraction, and the generated LARs can be automatically uploaded on different application servers to optimize server and network loads.

There is no magic, anyway, and such a tool cannot invent missing information or features. When an interoperability project starts, one of the first actions should be to ensure that each organization has, or can get, procedures and tools at the right maturity level.

When this condition is fulfilled, a semi-automatically generated LAR can do the job of representing the legacy application transparently.

References

1. Wiederhold, G.: Mediators in the architecture of future information systems. IEEE Comput. Mag. **25**(3), 38–49 (1992)
2. Hull, R., King, R.: Reference architecture for the intelligent integration of information. ARPA Technical report (1995)
3. Gottschalk, P.: Maturity levels for interoperability in digital government. Elsevier (2008)
4. Guettier, C., Yelloz, J., Lefebvre, A., Ponthoreau, P., Martinet, C.: Improving tactical capabilities with netcentric systems: the Phoenix'08 experimentation. In: MILCOM 2009. IEEE (2010)
5. Guettier, C., Yelloz, J., Cherrier, O., Mayk, I., Lamal, W.: Interoperable joint planning and execution web service with TITAN. In: MILCOM 2011. IEEE (2012)

Recent Advances in Caller Localisation for Public Safety Answering Point

Biagio Lanziani(✉) and Michele Biolè

Regola s.r.l., c.so Turati 15/H, 10128 Turin, Italy
b.lanziani@regola.it

Abstract. Caller localisation is a key factor for the success of some emergency request to PSAP (Public Safety Answering Point). The article examines the state of the art of the caller localisation across several situations and technologies. In the end will be detailed a commercial and innovative solution in this field called FlagMii proposed by the Regola company that suits the localisation needs of PSAP when the caller use a mobile phone for a service request.

Keywords: PSAP · Localisation · Caller localisation · 112 · FlagMii · Regola

1 Introduction

If the person in need of help is the one calling, he/she can be in different situations; lost in a forest or on a mountain, in a sinking boat at sea, in a burning house or a crashed car, or having trouble to breathe, losing consciousness, be in severe pain or being robbed or beaten. The caller can be deaf, speaking a foreign language or being unable to speak at all. Regardless of this or kind of emergency, the PSAP must handle the case in a most efficient way, and the handset, and in general mobile devices, can be a great help in this.

In general every emergency request need to be localised, also the one where the caller declare his position by address.

It's important to stress that a PSAP is always striving after verified information. Only if a PSAP gets verified information we can send the right resources to the right place and help the right person(s). And also, be sure that an emergency really has occurred and that this is not a prank call. The two basic questions that the PSAPs needs a quick answer to when an emergency call is made are:

- WHAT has happened, and
- WHERE this has happened

If the mobile device or Mobile Network Operator can assist the PSAP with swift and correct data, for example a reliable location by GPS, this can shorten the interview and enable a quicker dispatch of resources. Swiftness saves lives.

The same result occurs if the PSAP can resolve an address in its position in a reliable way.

R. Szewczyk and D. Havlik (eds.), *Recent Trends in Control and Sensor Systems in Emergency Management*, Advances in Intelligent Systems and Computing 675,
https://doi.org/10.1007/978-3-319-70452-4_5

Every piece of information that can help is also of benefit, for example the name of the owner of the mobile, language spoken, known diseases (in case of sickness), and of course the telephone number [1].

The article, basically, treat the existing solution that can help the PSAP in the caller localisation when he/she is requesting an emergency service from mobile devices.

2 How Is Possible to Localise an Emergency Call

Every emergency call need to be localised, not only the call performed using a mobile device but is quite clear that to localise a call using a mobile device is more challenging. Indeed when the caller use a fixed line to call 122 the telephone has a fixed endpoint that the Telecommunication Company has, probably localised.

With the call performed by mobile devices the position is not known and need to be calculated/gathered in some way.

These mobile devices we consider can be:

- A mobile phone with or without geo-localisation equipment and with or without data connection.
- An IVS (In-Vehicle System) that perform an eCall. The IVS is the equipment who physically execute the emergency call in case of an eCall. It acts basically as a mobile phone with the advantage that the position of the call is sent to PSAP using a specific technique. The eCall and its localisation system is described in Sect. 3.1.

With every mobile device the caller ask for a service just dialling the emergency number, like 112. Under this condition the technological challenge is to try to localise the caller using information that can be gathered either from:

- The Mobile network: The mobile network knows, with a degree of accuracy the position of the caller.
- The Mobile device: the device can know its position using the information derived from the radio signal from the network and/or got from the equipment on board like GPS (Global positioning system) equipment.

2.1 Fixed Line Localisation

The location of calls coming from fixed telephony is based on data owned by telecommunication companies which has the geo-localised data for every number connected to the network.

As there are many companies, each of them must make its information accessible.

The following aspects should be taken into account:

- The centralisation or decentralisation of data coming from all telecommunication companies: in some countries a central database comprises data from subscribers from all companies, in other cases there are different databases, normally one for each company.
- Location of the data: the database or databases can be stored in the PSAP or not.

- How emergency services access data: if the database is not located in the PSAP it can be accessed remotely.
- Accuracy of the data:
  - How often they are updated: changes in subscribers' data may occur daily;
  - Standard format: it is necessary that all data respects the same structure;
  - Correctness of the address (national number portability included);
  - Availability of caller ID and address for calls coming from a campus network;
  - Existence of private numbers.

2.2 Mobile Network Operator Localisation

A mobile phone need to use a MNO GSM cell. The cell is the 'unit' who can physically manage the call and the Cell ID is the identity number associated with a cell, which is designated by the network operator. This information is used in the network during normal operation to identify the connection point of the mobile to the network. The operator knows the co-ordinates of each cell site and can therefore provide the approximate position of the connected mobile. The Cell ID positioning considers the location of the base station to be the location of the caller and communicates the sector information. The network cannot guarantee that the serving cell, which is used to estimate the handset location, is the closest to the caller. The accuracy of this method depends of the size of the cell. It can vary from a few meters in urban locations to 10 to 30 km especially in flat countryside and water surfaces. The underlying issue is that mobile phone networks are optimised for coverage, capacity and call handling with minimum number of cells rather than for locating phones. This method can be used regardless of the type of phone but the provided accuracy and reliability are not according to emergency services needs [2].

2.2.1 Cell ID with Timing Advance

The measured time between the start of a radio frame and the arrival of data to the cell of the mobile network can be added to the data of the cell identification. This period of time is called Timing Advance (TA). Information derived from the wireless network can also be incorporated to the Cell ID based method. This way accuracy can be improved.

2.2.2 Cell ID with Timing Advance and Received Signal Strength

Advanced systems determine the sector in which the mobile phone resides and estimate also approximately the distance to the base station. Further approximation is ensured by interpolating signals between adjacent antenna towers.

Qualified services may achieve a precision of down to 50 m in urban areas where mobile traffic and density of antenna towers (base stations) is sufficiently high. In rural and desolate areas base stations may be kilometers apart and therefore locations are determined less precisely (Fig. 1).

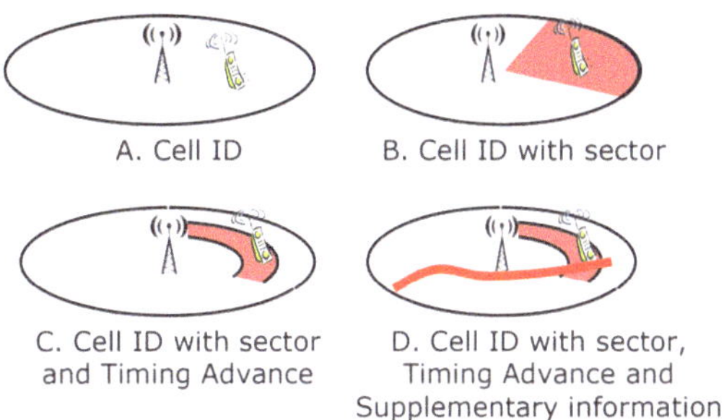

Fig. 1. Cell ID

2.2.3 RF Pattern Matching Method

The RF Pattern Matching technology is based on the observation that the radio environment and signal strength varies from location to location due to features such as terrain, buildings and cellular signal coverage. If enough elements of the radio environment can be measured with sufficient accuracy, each set of measured values provides a radio signature that uniquely identifies a particular location. RF Pattern Matching can provide high accuracy location information.

2.2.4 Uplink Measurement Methods

Here some examples of the available uplink measurement methods:

- Time of Arrival Method (ToA): This technology uses the absolute time of arrival at a certain base station. The time of arrival (ToA) is the travel time of a radio signal from a single transmitter to a remote single receiver. The time is a measure for the distance between transmitter and receiver. Time of arrival data from two base stations will narrow a position to two circles and data from a third base station is required to resolve the precise position with the third circle when matching in a single point.
- Angle of Arrival (AoA): The angle of arrival mechanism locates the mobile phone at the point where the lines along the angles from each base station intersect. AoA (Angle of Arrival) requires specialised receivers at the base stations in addition to the construction of directional antenna arrays on the existing cell towers.
- Uplink-Time Difference of Arrival (U-TDOA): It is a real time locating technology for mobile phone networks that uses multilateration (hyperbolic positioning) based on timing of received signals. Location Measurement Units (LMUs) are co-located at the Base Transceiver Stations (BTSs) to calculate the time difference measurements used to determine the location of a mobile phone. The technique is a network-based location technology, so it can locate any phone. It can also locate any phone in any environment – including indoors and in urban areas with tall buildings. The accuracy is within 50 m. Typically, the time to first fix is about 6 or 7 s in GSM and about 10 or 11 s for UMTS.

2.2.5 Downlink Measurement Technologies

- OTDOA-IPDL (Observed Time Difference Of Arrival - Idle Period Downlink) Method with network configurable idle periods: The OTDOA-IPDL method involves measurements made by the user equipment and Location Measurement Unit of the UTRAN (Universal Terrestrial Radio Access Network) frame timing e.g. System Frame Number (SFN) observed time difference). These measures are then sent to the Serving Radio Network Controller (SRNC) while, in networks which include an Stand-Alone Serving Mobile Location Center (SAS), they may be forwarded to the SAS. Depending on the configuration of the network, the position of the User Equipment (UE) (mobile phone) is calculated in the SRNC or in the SAS.

The material reported here in this section provide a general overview of the technique the MNO can use to get the caller position using the radio link between the cell and the mobile phone. At the end, from the PSAP point of view the information the MNO can provide are depicted in the following Table 1:

Table 1. MNO information

Field	Description
Number	Phone number to be localized, fully qualified by the international area code
ResultCode	Localization status like: 0: localized −1: unauthorized −2: the number is missed from the database −3: the number is in the database but it is not localized
IsMobile	True if the number is a mobile phone number
Timestamp	Date and hour of the position in seconds from the midnight 1/1/1970 UTC (it is the Unix time stamp)
Latitude	Latitude
Longitudine	Longitude
Accuracy	Localization accuracy in meters
Azimut	Localization azimuth (center of 120° arc)
Country	Holder's country
Region	Holder's region
City	Holder's city
ZipCode	Holder's zip code
Street	Rue/boulevard
HouseNumber	House number
DoorNumber	Door number
Name	Holder's name
Surname	Holder's surname
CardID	Holder's CardID

2.3 Mobile Phone Localisation

The mobile phone is in condition to determine its position and it can use the following technique [2].

2.3.1 Downlink Measurement Technologies

- Enhanced Observed Time Difference (E-OTD): The location is estimated using measurements made by the mobile phone, rather than by the base station. The location method works by multilateration.
- OTDOA-IPDL (Observed Time Difference Of Arrival - Idle Period Downlink). This method was already described in Sect. 2.2.5. The mobile phone may also perform the calculation of the position using measurements and assistance data.

2.3.2 Global Navigation Satellite System Based Technologies

- Assisted GPS (A-GPS): Standalone GPS operation uses radio signals from satellites alone. A-GPS additionally uses network resources to locate and also uses the satellites faster as well as better in poor signal conditions. In very poor signal conditions, for example in a city, these signals may suffer multipath propagation where signals bounce off buildings, or be weakened by passing through atmospheric conditions, walls or tree cover. When first turned on in these conditions, some standalone GPS navigation devices may not be able to work out a position due to the fragmentary signal, rendering them unable to function until a clear signal can be received continuously. In the case of mobile phones, if a GPS signal cannot be received, or if the handset does not contain an G-GPS chip, a fall back to network based location methods is required.
- A-GPS SIM: is a hybrid positioning solution comprising A-GPS, GPS, RF Pattern and Cell ID methods to ensure performance both outdoors and indoors. The assisted GPS receiver module is embedded in a standard size SIM card for legacy, new GSM and 3G phones. For retrieving the assistance data from the server and for transmitting the location data to PSAP, A-GPS SIM can use GPRS, like A-GPS phones do, but also USSD and SMS to support mid tier and low tier phone models. Although USSD and SMS are not as fast as GPRS, those can be used concurrently with a voice call, which is important in case of emergency calls. No software or hardware modifications are needed for the phones and no or only minimal modifications for the network. A-GPS SIM supports both automatic transmission of location information to PSAP when 112 is called and PSAP or network initiated requests. The smart card security features of the SIM can be used to encrypt the location data and to prevent unauthorised tracking of citizens.
- Hybrid positioning systems: are systems for finding the location of a mobile device using several different positioning technologies. Usually GPS (Global Positioning System) is one major component of such systems, combined with cell tower signals, wireless internet signals or local positioning systems. These systems are specifically designed to overcome the limitations of GPS, which is very exact in open areas, but works poorly indoors or between tall buildings. Wi-Fi signals may give very exact

positioning, but only in urban areas with high Wi-Fi density - and depend on a comprehensive database of Wi- Fi access points. There are situations where A-GPS could fall back to another high accuracy location technology like U-TDOA. In fact, in optimal situations where A-GPS and U-TDOA can work, both location technologies can be employed, and the calculations can be combined to offer location accuracies superior to either technologies working individually.

Other Technologies

Indoor proximity detection: where operators deployed indoor coverage infrastructure using distributed antenna systems (DAS), it is possible to add "proximity sensors" which can identify which DAS port a device is using and isolate it's potential location to a specific part of the building (office, airport, stadium). This provides 100% indoor yield to GPS type accuracy in environments where GPS often fails to work.

The mobile phone can discover its position and the accuracy of this position using the technique described above. The challenge is to transfer this information to the PSAP in some way.

3 PSAP Caller Localisation Technique

3.1 eCall Localisation Technique

In the near future (April 2018), every new car will have an electronic safety system automatically calling emergency services in case of a serious accident. Even if you are unconscious, the system will inform rescue workers of the crash site's exact whereabouts, and the rescues will be on its way within minutes. The system, which has been baptised "eCall", is going to work all over the European Union. It will soon be rolled out across the EU plus Iceland, Norway and Switzerland.

As soon as the eCall device (the In-veicle system) in the car senses a severe impact in an accident, it automatically initiates a 112 emergency call to the nearest emergency centre and transmits it the exact geographic location of the accident scene and other data. With the same effect, eCalls can also be made manually, at the push of a button (Fig. 2).

This is convenient if, for instance, you become witness of an accident. Whether the call is made manually or automatically, there will always be a voice connection between the vehicle and the emergency call centre in addition to the automatic data link. This way, any car occupant capable of answering questions can provide the call centre with additional details of the accident [3].

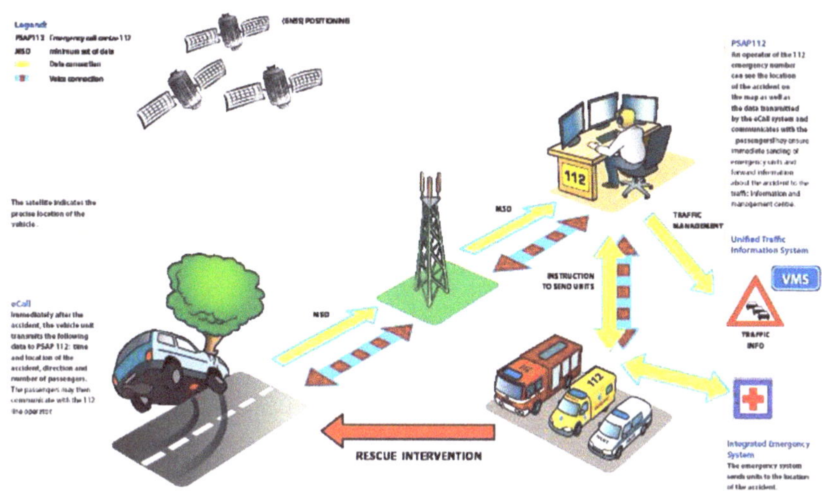

Fig. 2. Overview of the eCall

3.1.1 How PSAP Can Receive Position of a eCall

A PSAP to manage an eCall should be technologically enabled with (Fig. 3):

- Dedicate lines to receive eCall
- An in-band Modem necessary to receive the position of the vehicle in the, so called, MSD (minimum set of data)

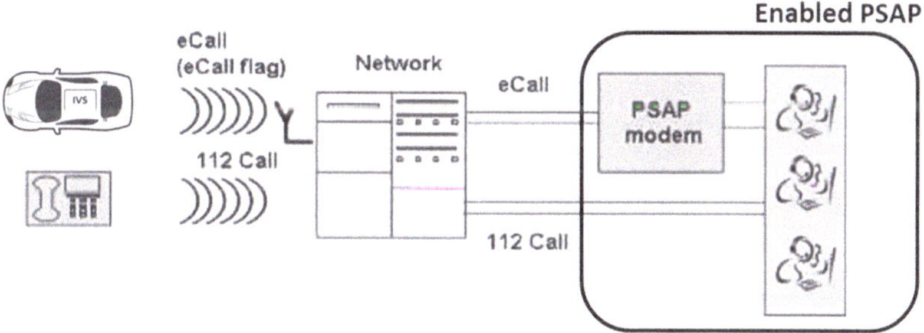

Fig. 3. eCall enabled PSAP and eCall

The solution works as follow [4]:

- The IVS execute a eCall. When doing this it rise a 'eCall flag' on the GSM network
- The 'eCall flag' allow the network operator to forward the eCall to an eCall enabled PSAP
- The PSAP recognize the arrival on an eCall and connect the IVS with an 'in-band modem' that allow the IVS to send the MSD. Within the MSD is contained the position of the vehicle.

Finally the content of the MSD (minimum set of data) is depicted in the Table 2 below. The data in bold represent the information about the localisation of the vehicle.

Table 2. The content of MDS

Data	Description
Activation	Whether the eCall has been manually or automatically generated
Call type	Whether the eCall is real emergency or test call
Vehicle type	Passenger vehicle, buses and coaches, light commercial vehicles, heavy duty, vehicles, motorcycles
VIN	Vehicle identification number (VIN)
Vehicle propulsion storage type	This is important particularly relating to fire risk and electrical power source issues (e.g. Gasoline tank, Diesel tank, Compressed natural gas (CNG), etc.)
Time stamp	Timestamp of incident event
Vehicle location	Determined by the on-board system at the time of message generation. It is the last known vehicle's position (latitude and longitude)
Confidence in position	This bit is to be set to "Low confidence in position" if the position is not within the limits of ±150 m with 95% confidence
Direction	Helpful to determine the carriageway vehicle was using at the moment of the incident
Recent vehicle location n (Optional)	Vehicle's position in (n−1) and (n−2)
Number of passengers (Optional)	Number of fastened seatbelts
Optional additional data (Optional)	MSD (at the vehicle manufacturer discretion)

3.2 Call Localised by the Network

In Sects. 2.1 and 2.2 we discovered that the network operator can localise the caller in any case, either when he call from a terrestrial line or when he call using a mobile device.

The PSAP can retrieve the localisation of the caller through a specific query to the network operator system.

3.3 Call Localised Using the Mobile Phone

EENA analysed several typologies of solution based on smartphone app. Here the brief summary of this research [5].

The solution presented here can include components deployed on the mobile phone with or without a server component. The article use the generic term of MSD (minimum set of data) to speak about the data set that contains also the position of the caller.

3.3.1 MSD Data Sent by SMS

In this solution, the information is delivered to the PSAP by legacy method (SMS – Short Message Service). The application will build the MSD without optional data and send it to the PSAP using SMS. The main problem in the SMS based solution is to know the most appropriate PSAP's number to send the MSD. One approach to solve this problem could be that the PSAP, after having received the 112 call sends an SMS to the caller with an e.164 number where the App can send the MSD. The MSD will be then sent to the PSAP (Fig. 4).

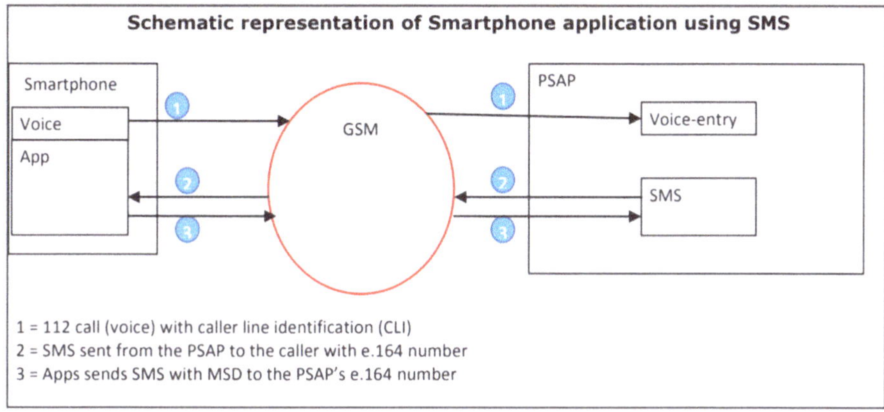

Fig. 4. Send MSD using SMS

It is worth mentioning that some European countries have already implemented e-SMS to contact to emergency services. As already mentioned, the strongest point of this solution is the already available SMS reception and dispatch technology in some PSAPs. The weaker sides are the delays that SMS can suffer and the restrictions of numbers of characters to be sent by SMS.

3.3.2 MSD Delivered Through Mobile Data Services - Centralised Server (Pull Option)

The Apps MSD should be sent using the data connection of the mobile phone and received by a server. PSAPs will connect to the central server and pull the MSD using, for instance, the caller line identification number of the caller (Fig. 5).

The main problems of this architecture are the not currently existing legal framework for sharing the data with the PSAPs and how to secure the transfer and (temporary) storage of the shared data. The positives sides of this architecture are the small update efforts to be made by the PSAPs and the flexibility of using the mobile data network to send the MSD (for instance format and number of characters restrictions).

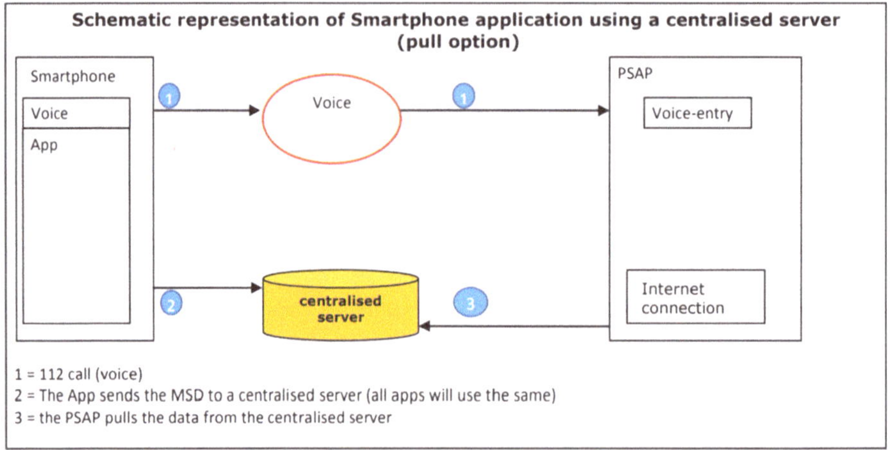

Fig. 5. MSD gathered by PSAP through centralized server – pull method

3.3.3 MSD Delivered Through Mobile Data Services - Centralised Server (Push Option)

The Apps MSD should be sent using the data connection of the mobile phone and received by a server that could then route data to the most appropriate PSAP (Fig. 6).

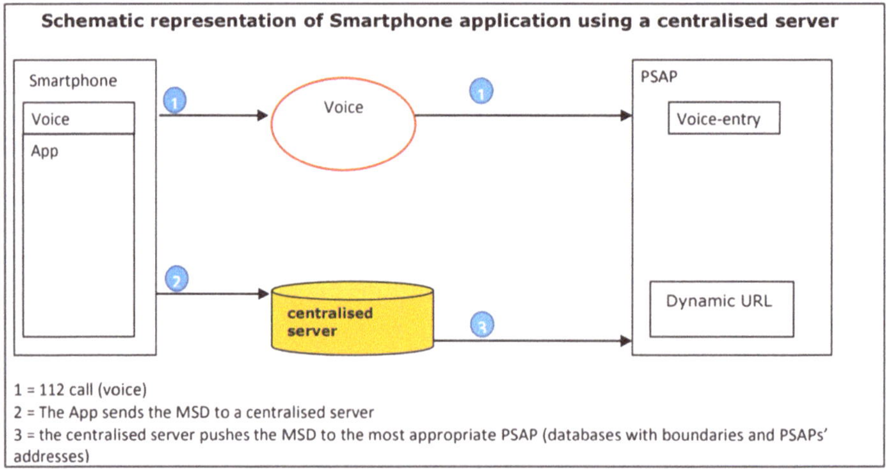

Fig. 6. MSD gathered by PSAP through centralized server – push method

This solution is compatible with future evolutions of technology. The PSAP receives the data with no delays in comparison with other architectures. One of the challenges of this solution is to manage the database containing the dynamic URLs (uniform resource locator) of the PSAPs and their boundaries. These data are currently not available. Subsequently, the database should be maintained.

As for the previous solution, the main problems of this architecture are the not currently existing legal framework for sharing the data with the PSAPs and how to secure the transfer and (temporary) storage of the shared data.

3.3.4 Combination of Mobile Data Services and SMS Approach

Callers make a 112 call to the PSAP. Consequently, the call is routed to the most appropriate PSAP. Caller location identifier and cell identifier (depending on the country) are available with the call. Once the call is received by the PSAP, the PSAP sends an SMS to the phone including its dynamic URL. The App can then send the MSD to this dynamic URL and it is received by the PSAP (Fig. 7).

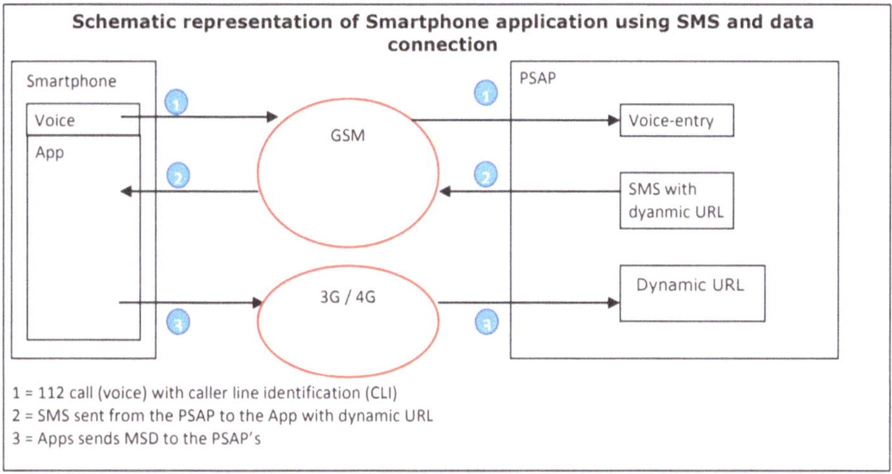

Fig. 7. PSAP gather MSD through SMS and data connection

In this architecture there is no centralised server, consequently problems described in the previous sections would not apply. Nevertheless, weakness of SMS has been already explained.

3.3.5 PSAPs Database and Boundaries Pre-configured in the App

The 112 application is preconfigured to send location data to a dynamic PSAP's URL (Fig. 8).

This solution is the easiest way to start with. One of the challenges of this solution is to manage the database. The first step would be to create a database with all dynamic PSAPs' URL. These data are currently not available. Subsequently, the database should be maintained.

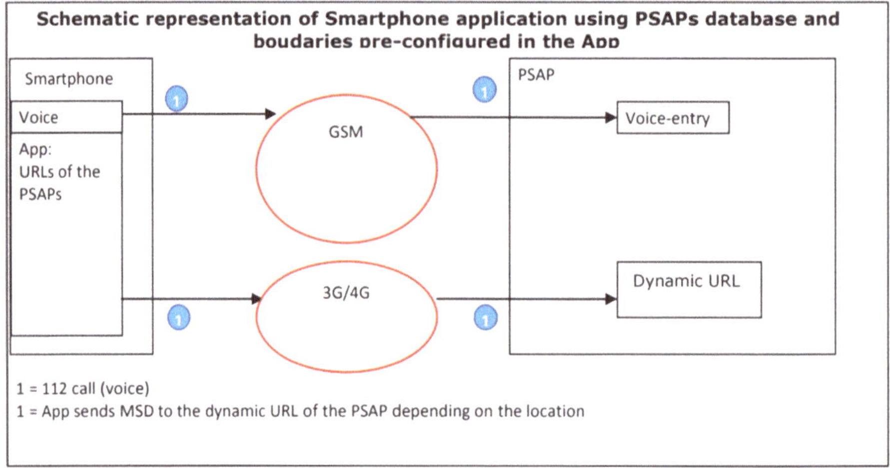

Fig. 8. 112 application

3.3.6 AML Advanced Mobile Location

The Advanced Mobile Location solution consist in a specific application on the mobile phone installed on the device [6].

When an AML-enabled smartphone recognises that an emergency call is made, it activates the phone's location services, including GNSS and Wi-Fi, and sends the caller's location information to the emergency services (to an SMS or HTTPS endpoint) automatically before turning the location services off again (if necessary) (Fig. 9).

The data path for AML data is shown below.

The success factor is to enable to AML as many devices as possible directly at the level of operating system, if possible. If not possible dedicated application installed after the purchase are good but there is the not negligible problem that the application must be installed before the execution of the call.

For this reason google decided to support this technology from all Androind o.s. from version 2.3. See the link:

- GOOGLE EUROPE BLOG: https://www.blog.google/topics/google-europe/helping-emergency-services-find-you/
- EENA http://www.eena.org/press-releases/aml-in-android#.WW92S4jyhEY

To complete the presentation of this technology here is depicted the structure of the SMS to be sent to the 112(999) number by the mobile phone (Fig. 10).

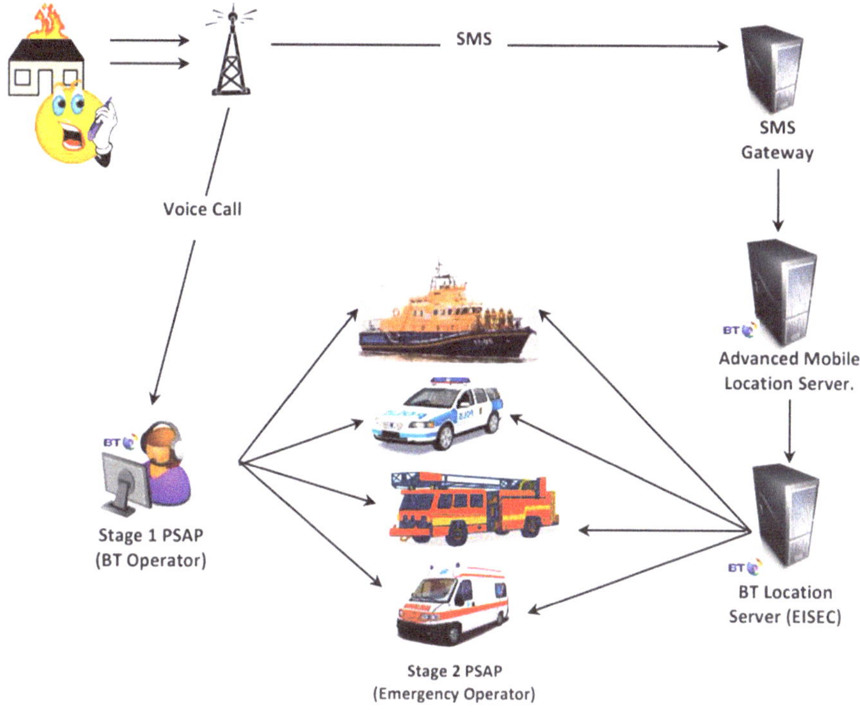

Fig. 9. AML implementation in UK

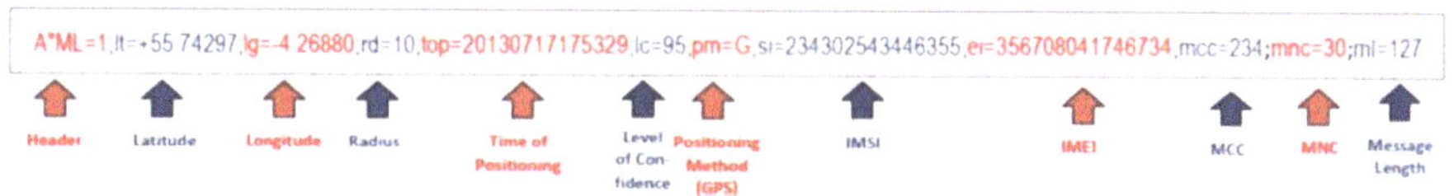

Fig. 10. Structure of the SMS

4 FlagMii Localisation Platform

FlagMii platform is a commercial example, produced by Regola (www.regola.it) who ensuring the Client to be able to locate rapidly those users placing an emergency call and requiring emergency response, as well as to perform enhanced functionalities such as Call Tracking, Textual CHAT, Picture Sharing, and more, in a bi-directional secure environment.

FlagMii Platform is based on 3 main components:

- A **Web Portal**, through which the Client can verify and see displayed on a map the location of a Caller. Information displayed contain approximate address, exact coordinates, location accuracy, last valid detection, etc.

Furthermore the Call Taking personnel, at the Client, can interact with the Caller by accessing to a specific CHAT functionality which allows a bi-directional real-time chat exchange.

- A free multi-lingual **Mobile App** for smartphones and mobile devices, available for iOS/Android/Windows Phone, that any User can early download and make use of all the available features when placing a traditional emergency call;
- A secure dynamic **Web App** being an excellent solution to Locate and Chat with Users placing a traditional emergency call but not having the Mobile App installed in their smartphones. Such solution does not require any download, and the Web App is reachable by tapping on a short URL contained in a special SMS, to be manually sent by the Call Taking personnel to the Caller's mobile device during the Call.

The operation of the system is shown schematically in the following figures (Figs. 11 and 12):

Scenario 1 - FlagMii Mobile App
Scenario 2 - FlagMii Web App

Fig. 11. FlagMii mobile app

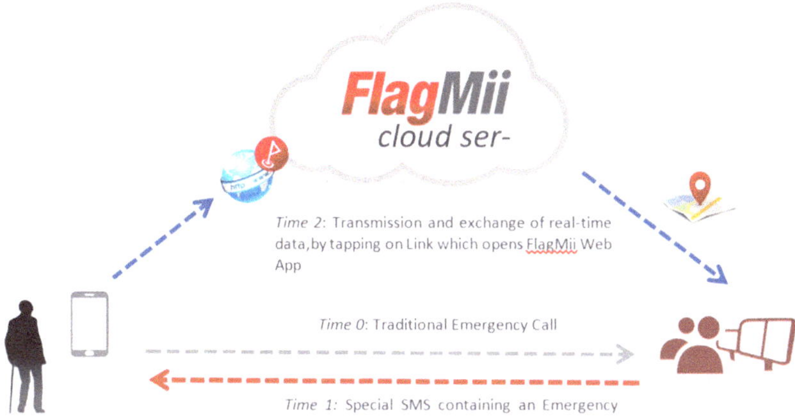

Fig. 12. FlagMii web app

4.1 FlagMii App Description

FlagMii Emergency Mobile System is a scalable and innovative service thanks to which Citizens, in case of need, can trigger an emergency call and establish active communication with the 911 or 112 Emergency Services, by performing simple operations on their smartphone or mobile device (Fig. 13).

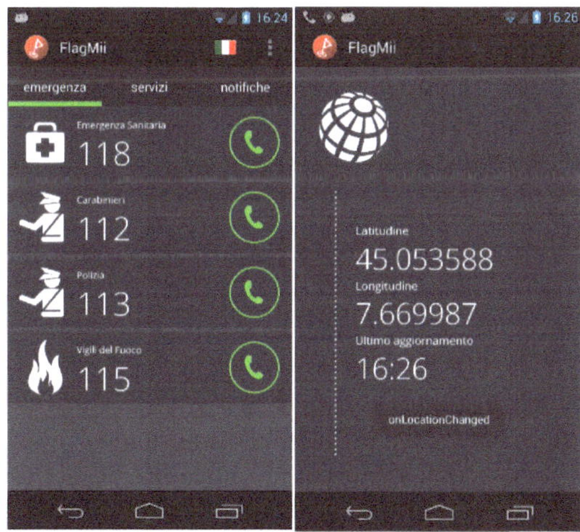

Fig. 13. FlagMii app description

Upon activating the call, where FlagMii Mobile App is available, this will push automatic location data from the caller to the cloud infrastructure of the Platform, making these data available on the Web Portal, on which the end-users can access and see displayed on a map.

The locating process will make use of all the smartphone/device capabilities from which a User is calling (GPS, WiFi, 3G, etc.) and these will be sent periodically for a given time, even in case the same call cuts off suddenly.

The approach described here above ensures to obtain a degree of accuracy when locating a caller extremely precise (in the order of 15–20 m) and in very short times (on average 5 s. from the acquisition of the call). Unlike other methods of locating (such as those based on Cell-ID), the location performed with FlagMii Mobile App maximizes the capabilities of the caller's smartphone, facilitating a rapid and accurate emergency response.

In general FlagMii is designed to ensure the security and confidentiality of data.

The entire technological platform built for the cloud components is certified **ISO/IEC 27001:2005**, is compliant with the specifications **SSAE 16/ISAE 3402** and also **HIPAA/HITECH**.

The location data are encrypted according to complex not-reversible algorithms, as a comprehensive guarantee of privacy of the caller. No one, not even the technical staff of REGOLA, can access in any way to such information.

4.2 FlagMii Web-Portal and Web-App

In the case a caller does not have the FlagMii Mobile App installed in his smartphone/device, the Web Portal allows a manual sending (performed most likely by a Calltaker) of a special SMS thanks to which the caller can be located as well.

The flow works basically as defined in Sect. 3.3.4 (Combination of mobile data services and SMS approach) with some small differences.

The flow works basically in this way:

- Thanks to the web portal functions the operator send an SMS to the caller
- The SMS contains an html link (link dedicated for the PSAP) where resides a specific standard html 5 web-app
- The caller follow the link and authorize the webpage to read the device position
- The webpage send the device position to the server
- The PSAP operator see the position of the caller.

This model has the big advantage of working even with not enabled devices. The only pre-requisites is to have an active data connection that, nowadays, is present on most of the circulating devices.

When the user access to the FlagMii web-app it can use additional feature like the chat. Indeed the webapp allow the PSAP operator to communicate with the caller using text messages. The operator can select the message to send from a set of pre-defined questionnaire. The function is particularly useful in case of an emergency request where the voice connection is:

- Not available for technical reason, located either on the network or on the device (mic failure).
- Too loud environmental noise that compromise the communication, for example during a concert, fire show, traffic in tunnels, etc.
- Deaf people.

5 Conclusions

PSAP localisation of the caller is success key factor for the emergency management. This article depicts a full list of available technologies through which the PSAP can localise the caller. The full set of technologies available today can decrease dramatically the rate of the not localised call because practically every call can be localised. Indeed, we have that:

- If a call comes from a fixed line it is localised through the Network Operator that know the position of the telephone, see Sect. 2.1.
- If a call comes from a mobile phone it will be localised through several technique that provide different data precision, see Sects. 2.2 and 2.3.
- If a call comes from a car the eCall standard provide the localisation of the vehicle, Sect. 3.1.1.

The article depicts quite clearly the technique a PSAP can act today to localise a call. Finally, the article depicts the head function of the Regola's Flagmii solution which is currently used form PSAP in Europe and Asia.

References

1. EENA Operations Document - Mobile Handset Requirements Communication to Emergency services, section 1
2. EENA Operations Document - Caller Location in Support of Emergency Services, section 6.6
3. HeERO Project. http://www.heero-pilot.eu/view/en/ecall.html
4. See standards EN 16072 and EN 16062
5. EENA Operations Document - 112 Smartphones Apps
6. EENA Case Study Document - Advanced Mobile Location (AML) in the UK

Improvements and Enhancements of Profile-Based Approach in C2-SENSE Crisis Management Project

Paolo Fabbri(✉), Vincenzo Russo, and Andrea A. Sbarra

Lutech S.p.A., Via W.A. Mozart 47, 20093 Cologno Monzese, MI, Italy
{p.fabbri,v.russo,a.sbarra}@lutech.it

Abstract. Empowering Command and Control (C2) systems with a harmonized view of information originating from different source providers (like physical sensors, messaging clients, legacy applications) is one of the key aspects of the C2-SENSE project: the effective management of emergencies is heavily dependent on information availability, its timeliness, its possibility to be correctly understood by specific consumers.

Achieving this goal in the highlighted context required to overcome some very specific and domain-oriented challenges, strongly influenced by the abundance of legacy architectures and applications found in the as-is, as well as their severely different technological level and maturity.

In this paper our intention is to show how, during the C2-Sense development, this service federation was made possible, which kind of issues had to be confronted with, which kind of solutions were evaluated and committed. In particular, we are going to delve into main aspects that can be improved, in order to suggest new investigation ways to make the project more declinable to real and complex crisis scenarios, giving a competitive advantage in being capable to reach an effective level of interoperability, making involved actors able to receive and understand information messages based on their actual informative value and not on the original capabilities of the convey systems.

In the conclusions we will also propose some further possible enhancements, built on top of the lesson learned during the C2-Sense research project.

Keywords: Interoperability · Legacy architectures · Sensor data · Stream management

1 Introduction

Effective management of emergencies depends on timely information availability, reliability and intelligibility. To achieve this, it is important to have a clear concept of preparedness under two different, but complementary point of view: response capability and management of measures to maintain operational continuity.

The project C2-SENSE considers as one of the most interested final users the civil protection organizations which aim at improving the quality of interventions, the recovery from the critical event and the capability of interacting with other actors involved in the rescue activities.

R. Szewczyk and D. Havlik (eds.), *Recent Trends in Control and Sensor Systems in Emergency Management*, Advances in Intelligent Systems and Computing 675,
https://doi.org/10.1007/978-3-319-70452-4_6

As Quarantelli states in his main works [1–5], it is very important to define a list of procedures to help focusing on the response capability and then be able to anticipate consequences of disasters. This allows to integrate the process of emergency management into a wider context embedded into involved companies and agencies.

One of the main point to address is to have a high-level interoperability approach, since it is important to allow heterogeneous private and public agencies/organizations collaborate each other in the most optimal way on both business processes and technical infrastructure.

One of the main results of the workshop I-CriMa for Interoperability and Crisis Management [6] is that it is acceptable to consider that in the next few years, organizational issues will be considered and solved, whereas IT issues could be considered.

However, unless standards and well-defined specifications are used, the interoperability of these systems can be very complex, since sensors standardized activities in the emergency domain, already widely used in several applications and contexts, are quite new.

The main objective of the C2-SENSE project is to try to overcome such fragmentation and provide a higher level of abstraction allowing a profile-based emergency interoperability framework using existing standards (different on per-case basis) and semantic-enriched web services to expose the functionality of C2 systems, sensor systems, and other emergency management/crisis functionalities.

For this reason, a "Profiling" approach was used to achieve interoperability by addressing all levels of the communication stack in the security field.

In the C2-SENSE project, the developed aspects relate to all the layers of interoperability stack chords shown in Fig. 1, implementing the missing technologies and making them available to the emergency community.

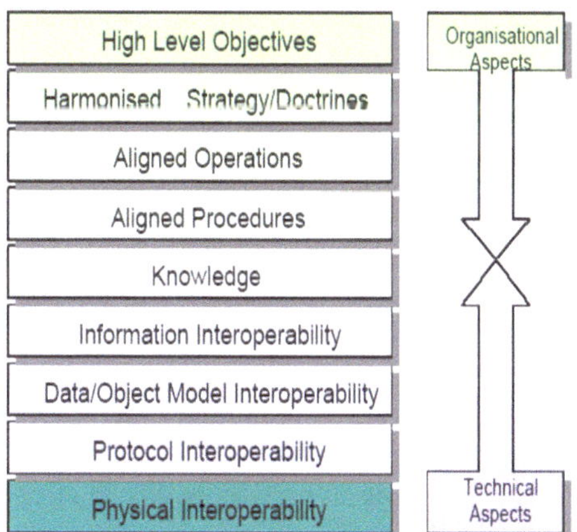

Fig. 1. Interoperability stack of the C2-SENSE project

Gaps in technology and standards are identified and one of the first complete implementations of the Interoperability Stack is implemented through the C2-SENSE Interoperability Framework.

2 C2-SENSE Architectural Pills

The approach depicted in the previous paragraph puts on the ground the need of a clear layered architecture in order to modularly address all technological and business constraints.

Figure 2 illustrates the overall architecture implemented in the project and depicts the approach followed in the C2-SENSE Emergency Interoperability Framework. Figure is quite self-explanatory, it structures the layers highlighting the main tools or devices that are involved. The quantified specific objectives of the project for each layer of the Interoperability Stack are as follows:

- **Physical Interoperability Layer**: Managing the physical connections between the networked applications and devices.
- **Protocol Interoperability Layer**: Managing end-to-end delivery of messages and documents.
- **Data/Object Model Interoperability Layer**: Managing data exchange among the disparate systems through common standard interfaces.
- **Semantic Information Interoperability Layer**: Managing provision of semantic mediation among different but overlapping common standard interfaces.
- **Knowledge Layer**: Managing creation of a common operational picture of the crisis situation and having the support of collaboration for joint decision-making.
- **Aligned Procedures and Operations Layer**: Managing alignment of emergency partners on their procedures and operations and reaching of an agreement.

A special remark has to be done for the *Interoperability Layer* which hosts an important component: the "Web Service Creator" tool. It represents the module that exposes the functionalities required by legacy systems. More in details, every component is implemented as web services which is conform to the developed Emergency Interoperability Profiles by creating all the necessary IT infrastructure through several open source tools.

Web services are customized according to the actor it is implemented for, using the specific requirements and methods. Web services communicate directly with an efficient Enterprise Service Bus implementation which orchestrates the communication between every actor involved in the data flow. It represents a messaging system that literally permits integration between different architectures, using interface adapters and data transformation services.

The abstraction based on the web services is possible thanks to several Protocol Adapters that overcome the heterogeneity in the protocols used by the Emergency applications/systems and sensors. Protocol Adapters (both standards-based or proprietary) connect to these systems/sensors and allow them to be accessible through web service protocols.

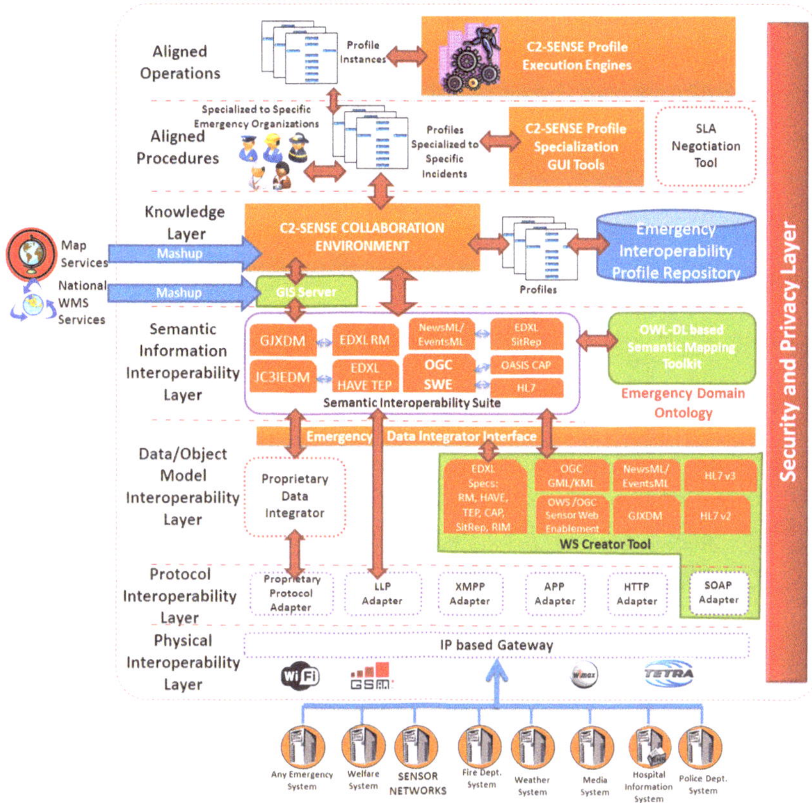

Fig. 2. C2-SENSE framework

3 Legal Application Representatives (LARs) as Abstraction

Legal Application Representatives can be considered as "*containers*", consisting of the Protocol Adapters, Data Integrators and Semantic Interoperability Suite from Protocol Interoperability, Data/Object Model Interoperability, and Semantic Information Interoperability layers, respectively. They literally represent the connection between C2-SENSE system and end-user systems. LARs handles this connection and provides communication among systems through an Enterprise Service Bus (ESB).

Figure 3 below gives a deeper detail of the LAR/ESB infrastructure.

As already said, C2-SENSE exposes services to applications and organizations such as "Web Services". For these services, interoperability profiles and corresponding Protocol Adapters have been developed.

These protocol adapters will be used to access emergency applications/systems and sensors in the C2-SENSE architecture through Web Service protocols.

There are two types of protocol adapters:

1. Sensor Protocol Adapters
2. Emergency Applications/Systems Protocol Adapters.

The first are basically the ones directly connected to the hardware of the sensors, which is ideally infinite, because of the specific sensor technical descriptions.

The second ones are based on the description of the structure of the messages and depend on the information really needed by each actor of the emergency communication.

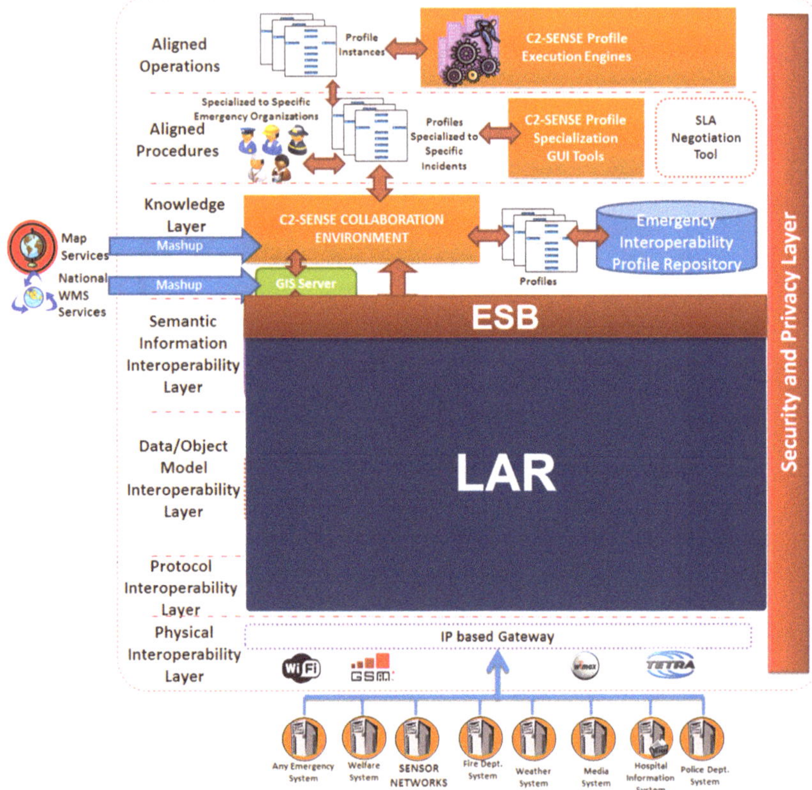

Fig. 3. Focus on the ESB and LAR infrastructure

4 Legacy Applications

Legacy applications can be very numerous and offer a large number of interfaces, standard or not, going from the simple file depot (FTP, SFTP, SFP, HTTP – all exposing CSV files, documents, workbook, and so on) to proprietary interfaces whose specifications aren't documented properly. On top of this, it should be added that some organizations offer no interfaces at all, or not machine-readable (fax, phone, bipedal transport, face-to-face meetings etc.) anyway.

In this kind of project will not know the list of all the protocols involved, for this reason a strategy has to be set up so that the system can be reasonably agile and capable to adapt easily to each different protocol.

When we analyze the long list of possibilities, we can identify the following groupings of protocols:

- REST: the system exposes stateless resources
- Services: the system exposes services with heterogeneous – generally stateful – technologies (SOAP, CORBA etc.)
- Diffusion: the system pushes data/updates on a given channel and expects no answers.

More complex behaviors can be found on top of these, but this concerns procedural interoperability, based on harmonized strategy doctrines.

Some protocols of interest to C2-SENSE pilot applications are presented in Table 1 below

Table 1. Protocols of interest to C2-SENSE pilot applications

Organization	Protocol	Group
Fire brigade	RSS/ATOM & CAP	Diffusion
Volunteers	TCP/Proprietary	Service or diffusion
Prefecture	SMTP	Diffusion
Municipality	ActOnline proprietary	Service
People	Facebook Twitter	Diffusion

5 Enterprise Service Bus: Performant and Agnostic Infrastructure

The Enterprise Service Bus has been described in [7] as "a new architecture that exploits Web services, messaging middleware, intelligent routing, and transformation". Since then, ESB technologies have become the subject of countless papers, works, and have generally known a widespread level of adoption across enterprise, SOA and Web-oriented contexts.

C2-SENSE ESB is based on the well-known Apache Kafka (Foundation, 2016), which is characterized by a high level of throughput, low-latency for handling real-time data feeds. The storage layer is a "massively scalable pub/sub message queue architected as a distributed transaction log".

C2SESB provides two main functionalities:

It exposes, through a WS-compliant interface, the publish/consume functionalities of the underlying service bus
It manages the logging of the operations executed on the underlying service bus.

Figure 4 illustrates a generic example of how a WS-compliant approach is employed in the C2SESB usage.

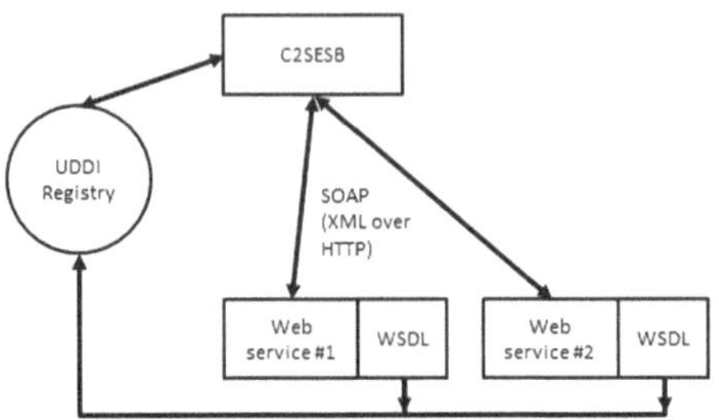

Fig. 4. C2SESB web services paradigm

Within each organization, programmers and system integrators develop Web Services that can discovery services, enquiry their capabilities, process their responses about their functionalities. The interface for each organization's Web Service is published using its WSDL and registered in a UDDI Registry.

Once the necessary information is acquired, the two WS are ready to interoperate.

6 Evolving the ESB Communication: From Static to Dynamic Approach

As it stands, C2Sense project is a solid pilot project that has a few possible short-term improvements that can potentially make a much more real-life aware solution or even evaluate properly the pilot effectiveness in a pilot-like real life usage sample.

The following paragraphs provide some topics that can be considered in order to review the design of some modules and processes, to gain efficiency and better exploit the great scalability potential available through the used technologies.

7 Evaluation Metrics for C2-SENSE

As it has been designed, C2-SENSE infrastructure seems to lack a sense of self-evaluation (Fig. 5).

Conceptually speaking, the whole project has the aim of increasing a sense of shared-awareness, allowing the actors to be in-real-time prepared to manage the crisis event and properly react.

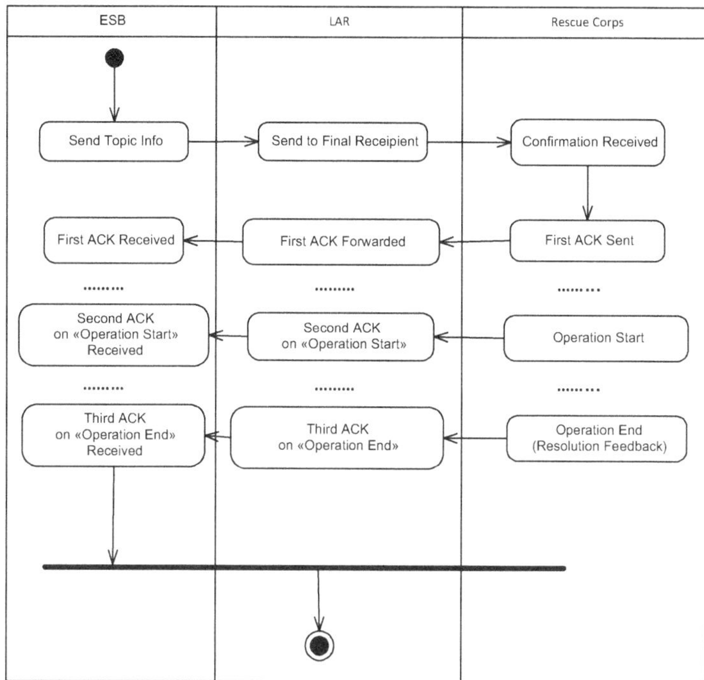

Fig. 5. The three-answer principle applied on the C2-SENSE communication flow

However, the system has no metric to evaluate its own impact. In fact it could improve the human handling (in terms of sheer "reaction time" and "resolution time") but will it? And if it will do, which will be the margin?

A first level of this shortcoming for a metric of evaluation, could be the introduction of a three answer principle. More in details:

- a subscriber upon receiving a message via the ESB, has to send a first message as an ack(-knowledgement) of the message,
- a second message relevant to the same topic as a "operation start" and
- a third and final message on the same topic will be interpreted as a "resolution feedback".

This would allow the system to track metrics of average resolution time at the beginning and the gradual decrease of the gap (in median), should there be an increase, as a testament of the effectiveness of the solution.

This is a raw solution though (as in a real solution two separate incidents don't necessarily have the same level of "priority", thus effectively a resolution of incidents should be accounted for the same median only if their priority is equal), alas, this sample workaround would be compliant with the current behaviour of the LARs, whereas introducing new fields would most likely require a change in some LARs.

8 Unstructured LARs Discontinuation

Some of the most antiquated LARs will prove a challenge to be properly integrated in the C2Sense landscape of integration, not for technical reasons, but for a purely semantic one: LARs that forward unstructured message body (e.g. email-like message) can be integrated with some workaround in the current setup of the ecosystem, but will prove more and more of an hinder, as they will most likely keep the system from moving forwards.

While from a technological point of view xml might seem a cumbersome format, heavy with context-overhead, said overhead allows xml to be descriptive and xsd structure or similar constraint allows a definite grammar to be paired with the communication itself.

9 Advanced Topic Subscription and Dynamic Profiling

As of now the topic subscription mechanism is working, but this is working based on some "glass bowl" level of isolation. The topic, that should have been something to ease the access to the information, have now become via the Process Engine a binding mechanism.

Regarding subscription there are two possible advancements:

- Geo-tagged topics: Basic information relevant to an incident happening in a certain city on a certain day should be shared and made available on a dynamic event-pipe, effectively for example this secondary message-pipe can be used to share//link an event that is relevant to the fireguard (e.g. intervention for apartment arson) with any potential other activity that happened the same day on the same city that involve different corpses. These message queue would be dynamic in content.
- Dynamic Profiles: Under the assumption that the topic geo-tagging is in place, we can, based on user feedback, have the system generate suggestion on how to tweak the Process Engine. For example if messages of a certain type relevant for a certain security assistance corp., when matching some specific tag value are acknowledged as relevant (again via some LARs provided feedback) by other corps, the system might suggest that the current Process Engine Profile for that type of message can be tweaked to alter the flow and alert secondary corps based on some conditional logic.

10 Application Management Enhancements

There are three basic area where the user experience of the C2Sense Ecosystem could be improved:

- *Topic Management and Profile Progress Console*: while not strictly necessary, a nice to have to monitor the system flow, is a console that will track the topic that are generated (search for older topics, base display is current day topics only) and their relevant progress, meaning viewing the profile execution engine operation on a specific topic via a graphical workflow-like representation.

- *General Volume Metric*: again not strictly necessary, but there are two aspects that haven't been "taken" into consideration. First of all a nice to have would be a metric detailing the amount of data (considering number of message and type of message) by each LAR.
- *LARS availability:* given that conceptually C2Sense is to improve the ongoing situation what's a staggering misstep is the lack of robust self-check routine. Meaning that each LAR should have the possibility to send a "standardised" message that is just an "existence pulse", meaning that the LAR is communicating that it's functioning properly. This behaviour is the basic is mission-critical systems that integrates multiple sources and allows to detect in a smart way when a specific source is not working properly: for example assuming that the agreement is that a LAR should send this "existence" message every hour, an alert can be communicated (via a console or via email alert to select recipient) if a LAR is not sending any "existence" message for more than 3 h, thus prompting a verification of correct configuration// status of the involved LAR.

11 Conclusions

The paper has reviewed the main aspects related to the C2-SENSE project architecture focusing on those that can be addressed by interesting improvements.

The main consideration is that the tool should increase the agnosticism and the capability to integrate as many sources as possible with the minimum effort on the development for adapting the different formats.

The data transmitted through the communication can be improved in quality by using some metadata stored in the proper format.

Suggestions have the final goal to define a concept of "Emergency event" that has to be adequately handled with the proper communication flow involving the "interested" actors.

References

1. Quarantelli, E.L.: Problems in Disaster Preparedness and Response, Preliminary paper # 106, Disaster Research Center, University of Delaware, Newark, DE (1986a)
2. Quarantelli, E.L.: Disaster Crisis Management, Preliminary paper # 113, Disaster Research Center, University of Delaware, Newark, DE (1986b)
3. Quarantelli, E.L.: Criteria for Evaluating Disaster Planning in an Urban Setting, Preliminary paper # 132, Disaster Research Center, University of Delaware, Newark, DE (1988)
4. Quarantelli, E.L.: Major Criteria for Judging Disaster Planning and Managing Their Applicability in Developing Societies, Preliminary paper # 268, Disaster Research Center, University of Delaware, Newark, DE (1998)
5. Quarantelli, E.L., Tierney, K.: Disaster Preparation Planning, Preliminary paper # 59, Disaster Research Center, University of Delaware, Newark, DE (1979)

6. Zelm, M., et al.: Interoperability for crisis management (I-CriMa). In: Enterprise Interoperability: I-ESA 2012 Proceedings (2012)
7. Schulte, W.R., Natis, Y.V.: "Service Oriented" Architectures, Part 1, (report from Gartner) (1996)
8. Keen, M., Acharya, A.: Patterns: Implementing an SOA Using an Enterprise Service Bus. IBM Redbooks, New York (2004)

Innovative Early Warning System for Natural Disasters – Case Study on Earthquakes with Earthquakeguard™ and NowTice™

Biagio Lanziani[1(✉)], Michele Biolè[1], and Giulio Delitala[2]

[1] Regola s.r.l., c.so Turati 15/H, 10128 Turin, Italy
b.lanziani@regola.it

[2] NowTech s.r.l., Via Caronda 172, 95128 Catania, Italy
giuliodelitala@gmail.com

Abstract. NowTech has realized Earthquakeguard™ a Rapid Alerting System of Seismic Events – Earthquake Early Warning System, based on a special sensor that does not generate fake alarms. Using this kind of sensors this innovative system is also able to provide a high precise and reliable structural monitoring system, furthermore these sensors are suitable to carry out seismic micro zonation studies and researches for hazard estimation.

REGOLA is a dynamic organization, developing fully in-house products with a modular approach. The ultimate purpose is to deliver comprehensive seamlessly-integrated software solutions and ready-to-use independent applications, keeping the whole technological framework open to external sources and standard integrations with 3rd party providers. Regola's Nowtice™ is the cloud service designed to send out critical warnings and simple notifications, thought for Organizations and Institutions who need to communicate effectively, timely and controlled, especially in case of emergency situations.

The article explains how these solutions can be combined in order to create an innovative early warning system focused on earthquakes.

Keywords: Earthquake · Early Warning System · EWS · Mass alerting system · Public warning system

1 Introduction

Public Warning Systems are needed to protect the lives of people in case of major emergency by warning the public of impending disasters. Tornados, tsunamis, hurricanes, floods, natural volcanic, and releases of deadly gas are examples of dangerous situations where PWS can save lives. Chemical plants, hydroelectric plants in dams and nuclear facilities are required to have the ability to notify the surrounding population of an industrial accident. There is no doubt that effective early warning systems have substantially reduced deaths and injuries from severe weather events [1].

This article introduces about the possibility that new technologies offer to extend this experience to earthquakes. Such events are unpredictable, wide and disruptive but

R. Szewczyk and D. Havlik (eds.), *Recent Trends in Control and Sensor Systems in Emergency Management*, Advances in Intelligent Systems and Computing 675,
https://doi.org/10.1007/978-3-319-70452-4_7

today exist the possibility to manage this events to build an early warning system who produces reliable alert with a significant advance of time that can be exploited to take some proactive actions and save lives. Moreover, this article describes in detail the characteristics of a modern mass alert system which is part of a public warning system. One of the best advantages of an integrated, modern mass alert system is the ability to provide people with **two-way communication**. Public warning systems of the past only spread emergency alerts, giving people information but neglecting the power of the environment to contribute valuable information.

Today's culture is much different. With social media, Internet Of Things (IOT) hardware and smart cities, people have become part of the stories they document, as they unfold. Regular citizens provide eye-witness accounts, IOT sensors provide live streams of data, as well as commentary, insight, and feedback. These ground-level accounts of situations not only inform the public of what's going on, but give first responders and public safety officials invaluable insight into the chronological order of events, the needs of the people involved, and a play-by-play of the event itself. This knowledge helps shape strategy, align proper response personnel, inform surrounding areas of potential danger, and better direct people in harm's way.

Muting these self-proclaimed reporters by removing their ability to respond to alerts seems archaic these days. People expect to be part of the conversation as well as part of the solution. Companies must provide this conversational, two-way dialogue in their emergency communication plan as much as for the company as for the employees. Mass alerts can include links to additional information and forums, provide a social media page where employees can upload pictures and videos, construct message boards in mobile apps, and much more to encourage participation.

In case of an **earthquake** what was mentioned before it is extremely true, thus a modern Early Warning System (EWS) should be connected to a new Public warning system capable to interconnect emergency agencies and the population. But the main challenge of Earthquake EWS is to avoid false alarms in order be fully accepted by the population who must trust it. In this document is described an innovative Earthquake EWS that can 'foresee' the arrival of the earthquake without generating false alarm. This EWS is connected with a platform that manage the mass alert communication using the new technologies and characteristic described above. The combination of these two technologies can help emergency manager to build a more proactive and efficient network able to the forecast and manage earthquakes.

2 EEWS – Earthquake Early Warning System - EarthQuakeGuard

NowTech is an Italian company which has its own corporate focus on environmental and structural monitoring systems. NowTech has realized an Earthquake Early Warning System (EEWS) named 'EarthQuakeGuard' that doesn't generate fake alarms.

The solution relies on distributed monitoring sensor network connected to a main remote server that collects and evaluate in real time the ground acceleration and ground motion of the area covered by the sensor network.

If a sensor detects a seismic event, it quickly sends information and data to the main server that checks if other network sensors have detected the same event before sending alert notifications to the enabled devices in the affected area. The double check procedure and the main features of the RSS-1 detectors (described later) ensure the absence of fake alarm notifications (Fig. 1).

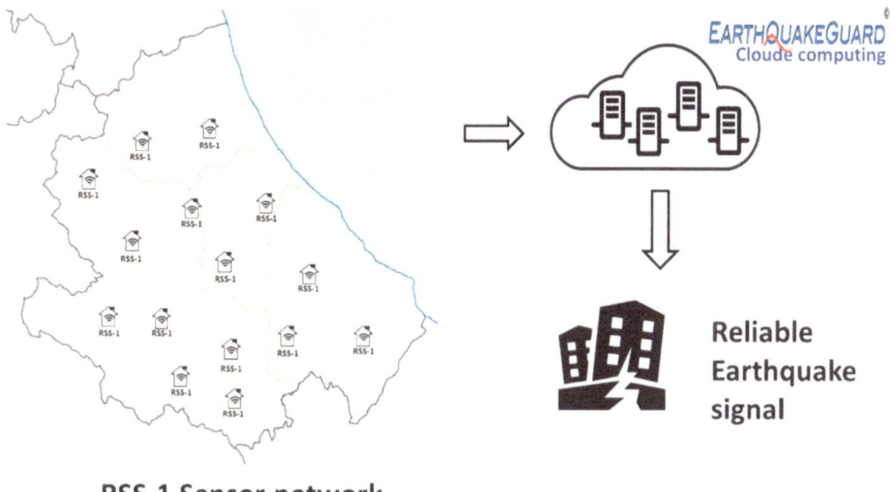

Fig. 1. Basic architecture of the solution

2.1 The Sensing and Monitoring Station RSS-1

The RSS–1 sensing and monitoring station is the sensor unit of this innovative EEWS. It can detect the full three-dimensional angular displacement (360°) of the structure on which is installed either in static and dynamic mode. This feature allows the sensor to be accurately trimmed on the structure where is installed hence it can generate very accurate tilt and twist information.

RSS–1 can detect angular displacements from a minimum variation of approximately 0.014° over all planes in the three dimensions, even obtaining the direction of the movements (Table 1).

Table 1. RSS-1 measures

Accelerometer	
Measures	Inclinometer
	Thermal (used for real time inclination measure compensation)
	PGA (Peak Ground Acceleration) PGV (Peak Ground Velocity) PGD (Peak Ground Displacement)

The minimum detectable acceleration is approximately 1 mm/s2, it is enough to sense the effects of any cracks, implosions or instant sagging of a pillar or slab; furthermore, the vertical axis monitoring can detect any subsidence or variation in altitude, hardly perceptible phenomena by traditional instruments. The RSS–1 is equipped by a seismic detector with the resolutions of the Table 2.

Table 2. Resolutions of RSS-1 sensor

Acceleration resolution	0,00025 g (2.45 mm/s^2)
Inclination resolution	0,014° (0.244 mm per meter of height)

2.1.1 Accelerometer Measurements

The three axes acceleration measures are acquired with 0.00025 g sensitivity and are dynamically performed. The embedded high-sensitivity digital accelerometers of RSS-1 allow the acquisition of very low acceleration values with excellent performance in terms of signal/noise ratio and error rate in the digitization process (14 bit of conversion resolution).

The digital band pass filter isolates and identify weak and strong signals due to natural elements such as wind and other atmospheric phenomena; the detector is also immune to anthropic elements that interact with the structure (Table 3).

Table 3. RSS-1 accelerometer specification

Maximum detectable acceleration	2 g
Maximum sampling rate	200 Hz (200 samples per second)
Acceleration resolution	0,00025 g (2.45 mm/s^2)
Inclination resolution	0,014° (0,244 mm per meter of height)
BPF (Band Pass Filter)	1,5 Hz–25 Hz
Sampling resolution	14 Bit

2.1.2 Inclination Measurements

Notwithstanding the usefulness of the dynamic measurements, to perform accurate diagnosis of a possible damage due to seismic stress, it is necessary the acquiring of static or slow dynamic measurements. The RSS-1 allows static displacement measurements.

The system compares the dynamic detected trends with inclinometers measures to provide the trim evolution of the monitored structure.

The embedded Inclinometers have a high sensitivity (0.014° sexagesimal) an excellent signal/noise ratio and a bandwidth that allows you to correctly identify the inclinations associated with the first vibration modes frequencies of the structure (few Hz).

The RSS–1 detector can provide an absolute inclination relative to the horizon, it assumes as a reference the first value acquired after the verification of the correct positioning and installation of the tool.

The inclination values are compensated by the embedded thermal sensor through detected temperature; it makes possible to monitor the variation of the tilt data to detect any abnormal results compared to the typical daily and seasonal temperature excursion; the compensations are usually never more than a hundredth of a degree.

2.1.3 Independent Tests

All the specification of RSS-1 reported in this article was verified by an independent centre specialized in Earthquake engineering [2].

2.2 RSS-1 Applications

This innovative sensor station allows you to realize a highly accurate detailed surveys and analysis on structure trim of the monitored building; this is really helpful for the structural safety and stability check. The RSS–1 allows a high accurate displacements description. Its characteristic, described before, ensure a substantial time advance on the drift of a hypothetical shift of the structure or building that manifests tendency to collapse.

2.2.1 Structural Monitoring

The structural monitoring is made by more than one RSS–1 detector installed in the same structure, both for single-storey buildings and multiple-storey buildings.

Usually this kind of application is realized installing multiple detectors in the same building but placing them in the radial and tangential direction of the structure; this configuration allows the comparing and combining of relative displacement and then the modelling of complex structure in a three-dimensional vision. The RSS-1 detector can be used for structural and trim monitoring of:

- single-storey buildings
- multiple-storey buildings
- bridges
- viaducts
- dams
- any other structure…

2.2.2 Seismic Monitoring and Seismic Micro-Zoning (PGA, PGV, PGD)

Starting from the recorded values RSS-1 can discriminate the seismic or micro-seismic activities through the evaluation of the following parameter: PGA - Peak Ground Acceleration, PGV - Peak Ground Velocity, PGD - Peak Ground Displacement. The estimation of PGA, PGV, PGD is done using the following measures:

- two measurements for mutually orthogonal horizontal components of acceleration, velocity or displacement
- one vertical measurement of acceleration, velocity or displacement

The acquiring of the real-time accelerogram based on the three axes data make possible the early identification of the Primary seismic wave (P-wave); the first seismic wave that occur after an earthquake with a prominent compression component. Any Earthquake rise 3 types of Seismic Waves [3, 4] (Fig. 2):

- P wave (Primary)
- S wave (Secondary)
- R/L waves (Surface waves)

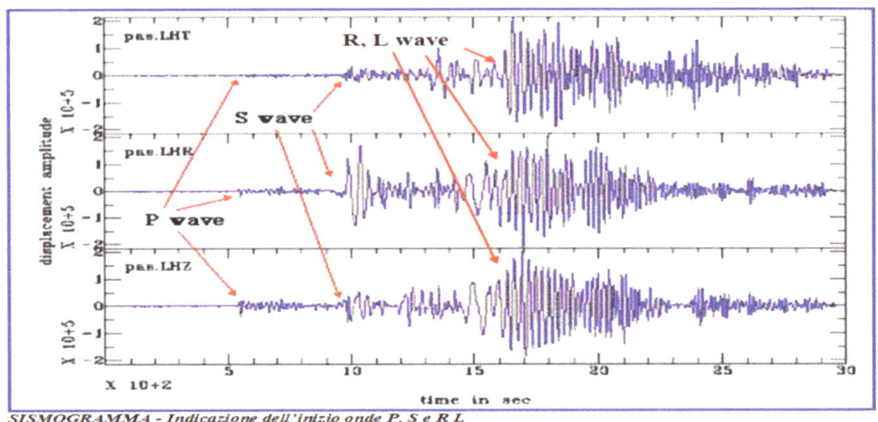

Fig. 2. Three types of seismic waves P, S, R/L

The P wave is the one that arrives first, more quickly than the others, but, fortunately, less dangerous since it is compression wave and usually it does not generate heavy damages.

The S wave travels about to half of the P Wave's speed and is a transverse wave and causes dangerous shakings.

The R/L Waves are the surface waves originated at the epicentre by a combination of P and S waves.

Surface waves travel along big distances before losing all the energy; R/L waves are the most damaging waves because generate vertical and horizontal oscillations of structures.

Fortunately, R/L Waves are slowest than the other Seismic Waves (about to 1/3 of the P-waves speed). Earthquake guard exploit this characteristic of the R/L waves to 'predict' the earthquake before it is felt by the population.

The PGA, PGV and PGD parameters are mutually independent, thus in case of an earthquake it is possible to have a high value of PGA and at the same time a low value of PGD and vice versa. The detection of maximum acceleration, maximum velocity and maximum displacement is useful but it is even more useful the estimation of time trends.

Build a Monitoring Sensor Network

To build an effective monitoring sensor network is necessary to install an adequate number of sensor in the territory who need to be monitored.

The minimal monitoring unit is composed by 2 sensors placed from 5 to 12 km of distance one to the other. This minimal unit can cover a territory of 80/100 Km2.

In order to increase the sensitivity of the network it is recommended to install more sensors. The finest sensitivity is reached installing sensors every 5 km.

The Fig. 3 show a placement project proposed to the governor of the Abruzzo region of Italy near the main city of 'Aquila' where exist a geological fault. The network cover an area of 625 Km^2 (25 km × 25 km) and use 16 sensors.

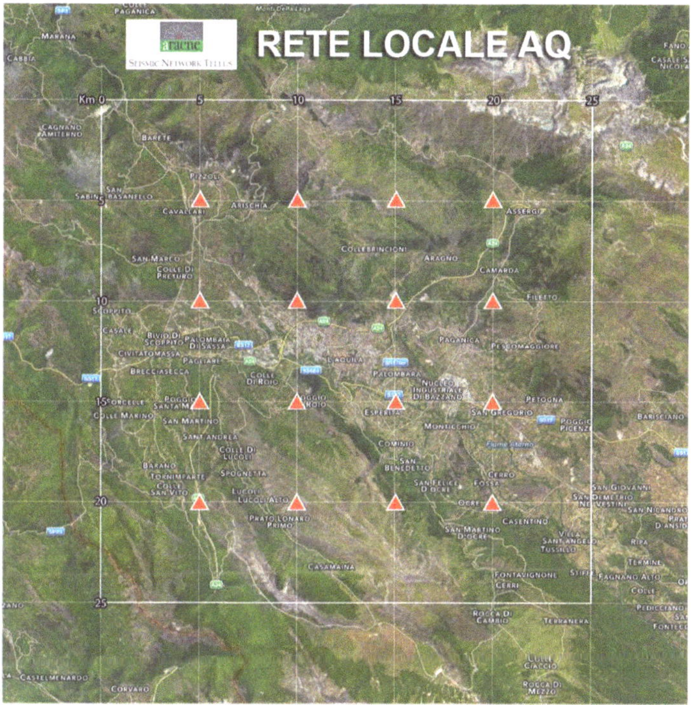

Fig. 3. RSS-1 placement project in Abruzzo region of Italy

In general, the density of the sensors should be higher on the geological fault, as depicted in the Fig. 3, and the sensor placed on a grid to promptly detect every seismic movement.

2.3 NowTech RSS-1 for EX-POST Analysis

The most widely used parameter to earthquake classification is the acceleration time trend, for example it is possible to identify single peak earthquakes or multiple acceleration peaks earthquakes, uniform distributed acceleration earthquakes or irregular distributed acceleration earthquakes.

From the analysis of the accelerogram is also possible the characterization of an earthquake according its duration and its strong-motion stage that can be generally identified by the time in which the acceleration exceeds the threshold value of 0.05 g.

The accelerometric data is useful to perform frequency analysis of the seismic event. An accelerometric dataset is collected through N samples in a T period, the maximum sampling rate ensure 200 samples per second.

The RSS-1 detector collect a dataset useful for frequency domain analysis, it allows you to calculate the Discrete Fourier Transform; the DFT analysis gives information related to the energy of the seismic wave and frequency distribution of peaks.

The following intensity scale (Table 4) incorporating the effects of PGA, PGV, PGD, duration of strong motion phase, and soil-structure interaction shows a correlation between instrumental intensity and potential damage of a seismic event.

Table 4. Intensity scale of earthquake

Instrumental intensity	Acceleration (g)	Velocity (cm/s)	Perceived shaking	Potential damage
I	<0.0017	<0.1	Not felt	None
II–III	0.0017–0.014	0.1–1.1	Weak	None
IV	0.014–0.039	1.1–3.4	Light	None
V	0.039–0.092	3.4–8.1	Moderate	Very light
VI	0.092–0.18	8.1–16	Strong	Light
VII	0.18–0.34	16–31	Very strong	Moderate
VIII	0.34–0.65	31–60	Severe	Moderate to heavy
IX	0.65–1.24	60–116	Violent	Heavy
X+	>1.24	>116	Extreme	Very heavy

3 Nowtice®

Nowtice is the application designed by Regola to send out critical warnings and simple notifications, thought for Organizations and Institutions who need to communicate effectively, timely and controlled, especially in case of emergency situations.

With a single click it's possible to spread any type of information via different medias/channels in order to reach timely the largest number of target recipients, accordingly to specific sending strategies configured and in general fully respecting the users' privacy.

Nowtice is a structured application to comply with any requirement for mass alerting, both for private and for public.

It can trigger different levels of alert by choosing different communication channels.

Nowtice implements a powerful driving force escalation that allows to apply different policies of retry and contact to ensure each recipient is timely notified.

Nowtice leaves the Client to define his own alert strategies with different severity, and for each of them to specify:

- The alerting channel to be used
- The escalation rules
- Target User and/or User Groups

The application allows to insert and maintain all the desired contacts and to collect themselves together in groups. The group management is highly flexible: it allows the definition in hierarchies to different levels, without limits, and it allows the association of a single contact in one or more groups, without duplicating unnecessary communications.

During the alert process the system User will be able to select one or more target groups and/or target contacts. Nowtice will generate a univocal list of resources to be reached and will manage simultaneous dispatches ensuring delivery (or a feedback of error in case of unexpected issue) and presenting real-time results in the application.

The platform allows the alerting of resources with/through various channels, some of these are traditional but at cost, while others are modern and above all free of any additional cost:

- EMAIL: Emails (with or without receipt) to the target recipient(s);
- SMS: SMS to the target recipient(s).
- VOICE: Voice Calls to the target recipient(s) using an automatic functionality named "Text to Speech".
- FAX: FAX to the target recipient(s) associated with the fax number.
- SOCIAL NETWORK: messages/notifications directly on the proprietary (of the Client) Pages or Profiles of the main Social Networks, such as Facebook, Twitter and soon Google+;
- MOBILE APPLICATION: push notifications on a specific APP for Smartphone and Tablet, exclusive and distributed free of any charge. This APP, called **FlagMii** is multi-platform and compatible with the major operating systems like iOS, Android and Windows Phone.
- WEB SITES, INTRANET;
- ROAD LIGHT SYSTEMS/PANELS, SMART TV;
- RADIO: message delivery to any resource equipped with radio terminals, both analog and digital.

This option foresees a radio network available and the technical availability (in terms of media licenses, drivers, SDK) for software integration;

This system operates in integration with any eventual pre-existing equipment and service, in addition to sensor systems and portable/vehicular radio.

Communication Channels

- PRIVATE CHANNELS: reserved and private communication channels only to a selected set of users.

A classic example might be delivering any potential notification/message to professional teams (or field-responders) with the aim to optimize and fasten the communication process.
With a single click the Client can reach the resources and keep in touch with any single person, with the certainty of delivery offered by certified protocols which was implemented.
This may result evident when rapidly retrieving the resources' availabilities and getting the resulting real-time situational overview in a special application area.
- PUBLIC CHANNELS: communication channels available for public use and targeted to population, with the aim to achieve a mass communication.

With a single click the Client can spread information towards multiple communication channels and timely reach the largest number of recipients.

For each generated alert the system will trace: operator who has created it, date and time, list of involved resources. In any moment it will be possible to look at the alert status, through a specific function, that will consider the following logical values:

- In creation: the alert is under creation;
- Scheduled: an alert being scheduled, which can be sent in different timeframes chosen by the user;
- Pending: the alert is in a Sending status;
- Completed: the forward of the alert was completed to all the recipients;
- Partial: the forward of the alert was completed only partially. The platform displays any single failed transmission, allowing a further retry or to cancel it.

The system will keep the history of the previous sent alerts, which will be available for consultation by the user (having appropriate credentials) in any moment.

For each of them it will be possible to see the level of escalation performed to contact any single resource.

As an example below few highlights on the current use that some clients have implemented:

- Public Safety pre-alerts for earthquake, hydrogeological, fire, ice and snow risks;
- Notifications for Closure of Schools or public places, regarding problems of street practicability;
- Institutional information or advices regarding Events and programs of public interest, being in progress;
- References and/or documents of public interest (weather forecasts, fire bulletins, etc.)
- Specialized Alerts to professional Teams or strategic resources to collect individual availabilities and replies in real time.

Below a brief summary of the points of excellence of Nowtice:

- No need to acquire locally a specific hardware or software;
- No assets to manage;
- CLOUD-based Service, with guaranteed High-Availability and H24 Support;
- Dedicated Web Portal for the use of location capabilities;

- Special SMS Sending for the location of users without the App (by referring to the Web App);
- Integration APIs to allow 3rd Party software providers to integrate.

3.1 FlagMii®

As mentioned above, Nowtice also includes integrated usability of FlagMii, which consists in the possibility to alert contacts with smartphones having the APP FlagMii installed in their devices. According the European Emergency Number Association (EENA) FlagMii is a **112 smartphone app** [5] free to use and available either for android or IOS devices.

Thanks to special features integrated with nowtice, "push" notifications will be used to deliver alerts on smartphones. Through FlagMii the user will be able to receive any alert or notification, to read it and also to access to further multimedia content of the alert.

Using the cloud will ensure the Client that every functionality is always available, making possible to locate a caller also without having FlagMii APP installed in the smartphone. By sending a text message with an URL, the caller can be located timely with particular accuracy facilitating the process of emergency response at the site.

FlagMii is designed for the public for a multiple use, among which:

(1) To locate exclusively in case of emergency call, displaying punctually on a cartographic map and in a dedicated portal with secure access;
(2) To receive alerts through the integratation with nowtice, via dedicated and/or customized channels.

In mission-critical contexts the displaying can be performed directly on the eventual pre-existing software system in use by the Client.

FlagMii is a multi-platform APP available for iOS, Android and Windows Phone operating systems.

FlagMii resolves one of most complicated problem in the emergency response, that is the correct identification of the incident location, especially in case of absence of references and panic status by the caller.

It's a certified APP from Emergency Authorities and compliant with the European Emergency Number 112.

FlagMii channel is provided free of charge because it's contextual to the use of nowtice.

3.2 Two Way Communication Capability Between Agency and Polulation

Today most part of the population uses intelligent mobile devices connected to internet. NowTice and his app Flagmii can really establish a two-communication channel among emergency agencies and the population. But Nowtice can do more and can manage also situation where the person who need to be rescued has not Flagmii app installed on the device. Indeed, with NowTice if the recipient has a simple internet connection then is possible to:

- Organize survey on the fly
- Get feedback any kind of feedback from the population

4 EarthQuakeGuard and Nowtice to Manage Seismic Events

The complete solution is composed by:

- A monitoring network of detectors NowTech RSS-1
- A cloud Server that collect data and rise events
- A network of 'Earthquakeguard actuators'
- The mass alerting system 'Nowtice'

With the process described in the previous section EarthQuakeguard generate a reliable earthquake signal to be exploited by Nowtice and a set of actuators (Fig. 4).

- Nowtice notifies people before the earthquake strong shaking arrives.
- Earthquakeguard actuators: is a set of devices who can take some proactive action that mitigate the risk of damage on the affected territory.

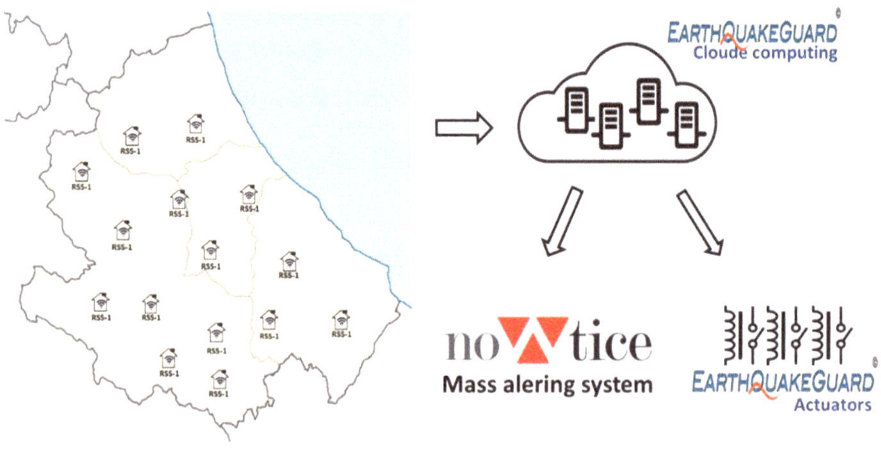

Fig. 4. EEWS complete architecture

Starting from the event detection, EarthQuakeGuard takes less than a second to quickly estimate the affected area, select involved users and send alert notifications through Nowtice (Table 5).

The entire solution has quantifiable capability to predict seismic events. Indeed, the disruptive S or R/L waves of an earthquake arrive later than the others wave (see Sect. 2.2.2), and this delay can be exploited to raise an alert before the earthquake is felt by the population.

The time of prediction (Is the period EarthQuakeguard rise a signal before the earthquake can be felt by the population) depends on three main factors: the distance between

Table 5. EarthQuakeguard response time

T0	Detection and sending time
T0 + 150 ms	Detection and sending time
T0 + 250 ms	Main server receiving
T0 + 400 ms	Double check, affected area identification and transmission time

the epicentre and the RSS-1 stations, the subsoil geomorphology and the hypo-central depth. Considering these variables, the solution can rise an earthquake alert **from 5 s up to 50 s**, before strong shaking arrives.

The solution gives you enough time to:

- Set emergency procedures
- Mitigate risks
- Take protective actions with the **EarthQuake guard actuators.** E.g. dangerous supply can be promptly stopped in case of earthquake)
- Ask people to reach a safer location (or to protect their self under tables or desk)

4.1 The Earthquakeguard Actuators

The Actuators are devices that can react to a seismic event raised by Earthquakeguard. There are several actuators ready-to-use like (Fig. 5):

- Visual and acoustic ALERT notification
- Electricity Grid Detach
- Gas Valve Detach and Block
- Automatic Doors Opening
- Hydro Valve Detach and Block
- Automatic elevators control
- Security Lights control
- Production Lines Releasing

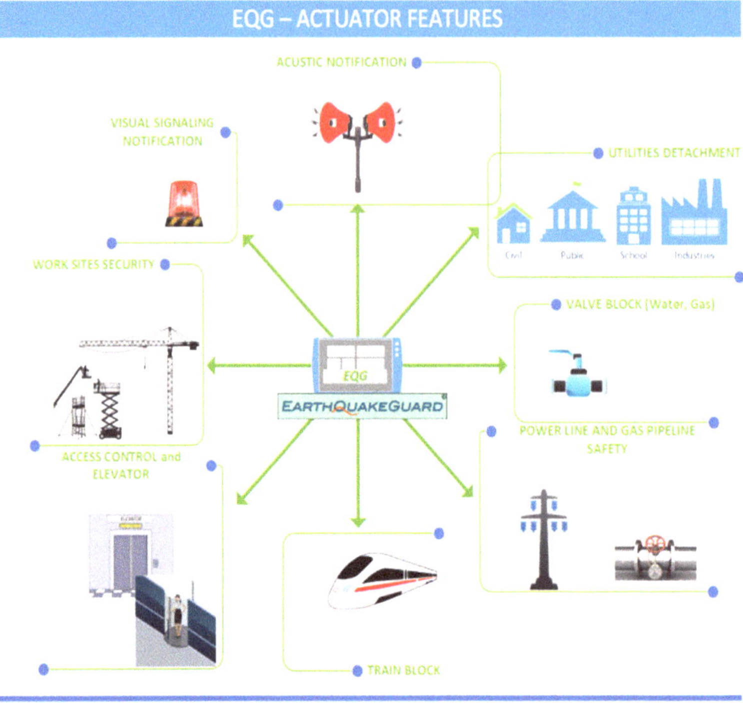

Fig. 5. EarthQuakeGuard actuators

5 Conclusions

Seismic events are still an unpredictable event (time - place - intensity) but this article demonstrates that today it is possible to deal with this event proactively, thanks to the combination of different technologies, with the aim of mitigating the damages resulting from a dangerous earthquake.

The core of the proposed solution is the construction of a seismic monitoring network, widespread in seismic-classified sites, based on NowTech RSS_1 detectors, which, as mentioned in previous sections/chapters, have the feature of not issuing false alarms because they measure uniquely ground acceleration and they are free from anthropic elements. This can be possible because the disruptive earthquake wave, the S and R/L one, arrive from 5 to 50 s. after the first one (see Sect. 2.2.2). This time can be exploited to raise an alert and to trigger a set of actuators (like valves, alarms, etc.) in order, for example, to close supply network or to alert persons who lives in buildings or factories. In other words, the usage of the innovative integrated EarthQuakeGuard/ Nowtice system allows you to safely secure any facility's buildings by suddenly acting on the solenoid valves, emitting an audible and visual alarm by means of sirens so that people can reach a less dangerous position even in the interior of the building. In parallel the system can send massively, using different channels and different criteria, messages who instantly alert Emergency Assistance Agencies, which will also be able to establish "bi-directional" communication with affected persons from the area affected by the seismic event. Earthquake Early Warning and communication between Emergency Management Agencies and the population is now possible thanks to the use of these innovative systems.

In the era of Smart Cities is not possible to ignore this great potential on earthquake management as well as the contribution that these tools can give to the protection and the safeguard of people.

References

1. EENA Operations Document – Public Warning. http://www.eena.org/uploads/gallery/files/pdf/2015_07_15_PWS_Final.pdf
2. EUCENTRE-European Centre for training and Research in Earthquake Engineering, Via Ferrata 1, 27100 Paiva, Italy. Document PROT. N. EUC316/212. http://www.eucentre.it
3. Aki, K., Richards, P.: Quantitative Seismology, 2 edn. University Science Books, Sausalito (2002). Dolce, M., Martelli, A., Panza, G.F.: Proteggersi dai terremoti: le moderne tecnologie e metodologie e la nuova normativa sismica, Milano, 21/mo secolo (2004)
4. Panza, G.F., Romanelli, F., Vaccari, F.: Seismic wave propagation in laterally heterogeneous anelastic media: theory and applications to seismic zonation. Adv. Geophys. **43**, 1–95 (2001)
5. EENA Operations Document – 112 Smartphone app. http://www.eena.org/uploads/gallery/files/operations_documents/2014_02_25_112smartphoneapps.pdf

Certification in Electronic Emergency Management

Mert Gencturk[1,2(✉)], Mustafa Sahin[1,2], Ezelsu Simsek[1,2], and Yildiray Kabak[1]

[1] SRDC Software Research & Development and Consultancy Corp., Ankara, Turkey
mert@srdc.com.tr

[2] Department of Computer Engineering, Middle East Technical University, Ankara, Turkey

Abstract. Like in any other domains, interoperability is a key challenge in electronic emergency management and it is only possible through standardization. There are a number of disparate standardization activities in this domain towards different layers of communication stack from physical interoperability to organizational interoperability. However, their extensive and effective adoption has not been achieved due to the lack of supporting certification mechanisms. In other words, claiming conformance to these standards is not formally described. In this paper, within the scope of European Commission supported C2-SENSE project, we propose standard mechanisms for the conformance/interoperability testing of the emergency management standards and their certification. The mechanism is based on CEN/WS Global Interoperability Test Bed initiative where an online test bed is accessed by the implementers of the emergency management systems, which claim conformance to the standards, and execute testing scenarios like in a real life setting.

Keywords: Conformance and interoperability testing · Certification · Test bed · Emergency management

1 Introduction

Software testing is the principal validation technique where the system developers test the individual components making up the system by using simulated test data. Each component is tested independently, without other system components. Certification, on the other hand, means that someone apart from the developer checks whether a component meets its specifications and has the desired functionality. In other words, a component is tested and certified that it has reached an acceptable quality standard before it is made available for the users [1].

Interoperability of a software or service with other systems is one of the most important indicators for the quality of software. Systems communicate with each other through software interfaces (i.e. protocols and documents) for which there exists many standard specifications in different domains. These specifications, however, include many detailed requirements that might be missed easily or misinterpreted during implementation. This might cause companies to lose money, time and even customers when their software is unable to interoperate with other systems in some business processes. In this

R. Szewczyk and D. Havlik (eds.), *Recent Trends in Control and Sensor Systems in Emergency Management*, Advances in Intelligent Systems and Computing 675,
https://doi.org/10.1007/978-3-319-70452-4_8

regard, certification of systems that are exchanging electronic documents between each other is crucial.

Certification has been implemented successfully in several domains such as eHealth and eBusiness. For instance, in 2008, Turkey's National Health Information System which was implemented based on HL7 v3 messages (HL7 Web Service Profile for transportation and HL7 v3 CDA Release 2 for the clinical document exchange [2]) has been certified by a generic testing tool named TestBATN [3]. Likewise, business to business (B2B) conformance and interoperability of UBL/NES (Universal Business Language/ Northern European Subset) based applications to UBL 2.0 specifications and NES profiles has been tested as part of Global eBusiness Interoperability Test Bed (GITB) Project [4] and the applications has been certified accordingly.

Although certification have been addressed successfully in several domains, to the best of our knowledge, it has not been addressed completely in emergency management domain yet. In emergency management, several command and control (C2) systems, sensor networks and civil protection organizations systems need to communicate with each other to manage the disasters effectively, hence interoperability of them is very important. These systems, however, use many different protocols and standard content models which creates a crucial interoperability problem. Therefore, it is necessary to test those systems properly and extensively after the integration. In this regard, within the scope of C2-SENSE project [5], we have implemented a certification mechanism in order to test the interoperability and conformance of different types of emergency management systems against the specifications defined in the C2-SENSE Emergency Interoperability Profiles. The system has been implemented based on the principles and specifications defined in GITB initiative.

The rest of the paper is organized as follows: Sect. 2 briefly explains the C2-SENSE project and the Interoperability Profiles defined in the project. Section 3 provides information about GITB project and the testing methodology. Section 4 explains the conformance testing mechanism implemented for the emergency management domain while Sect. 5 does the similar for interoperability testing. Finally, Sect. 6 concludes the paper.

2 C2-SENSE Project and Interoperability Profiles

C2-SENSE project's main objective is to develop a profile based Emergency Interoperability Framework by adapting the existing emergency domain standards and semantically enriched Web services to expose the functionalities of C2, sensor and other emergency management systems. In an emergency situation, several emergency organizations and systems communicate with each other to manage the disaster effectively. For example, in a flood case, different sensor measurements such as water level, temperature, wind speed etc. retrieved from different types of sensors are fused to create a common operational picture and shared among command and control systems. When a measurement exceeds some threshold value, an alert is broadcasted to inform corresponding emergency management organizations. In order to manage this kind of a situation effectively, the C2, sensor and other emergency management systems should cooperate flawlessly which would only be possible through interoperability and interoperability would only be possible through standards and well-defined specifications.

In this respect, a number of Emergency Interoperability Profiles meeting the requirements of emergency management have been defined in the C2-SENSE project as a collection of standard specifications covering all the layers of the interoperability stack [6].

The quantified specific objectives for each layer of the Interoperability Stack shown above are described in [7] as follows:

- Physical Interoperability Layer: managing the physical connections between the networked applications and devices. It addresses the physical connections such as GSM and WiFi.
- Protocol Interoperability Layer: managing end-to-end delivery of messages and documents. It addresses the transport level protocols such as SOAP, REST and SMTP.
- Data/Object Model Interoperability Layer: managing data exchange among the disparate systems through common standard interfaces. It addresses the XML based messaging standards [8] such as EDXL SitRep, EDXL RM, OASIS CAP and OGC SWE.
- Semantic Information Interoperability Layer: managing provision of semantic mediation among different but overlapping common standard interfaces.
- Knowledge Layer: managing creation of a common operational picture of the crisis situation and having the support of collaboration for joint decision making.
- Aligned Procedures and Operations Layer: managing alignment of emergency partners on their procedures and operations and reaching of an agreement.

Profiles define how to represent the information, how to send it through network or other mediums, and how to collaborate to manage emergency situation. The scope and usage purpose of C2-SENSE Emergency Interoperability Profiles can be summarized as follows. Further information can be found in [9].

- Alert: sending alerts to emergency management organizations and notifying them.
- Audit Trail and Node Authentication: providing centralized audit trail and node to node authentication to create a secured domain.
- Emergency Situation Map: providing common operational picture of the emergency area using a real-time data and geographical maps.
- Hospital Communication: managing hospital related information such as checking the number of available beds, location of an ambulance, availability of a service etc.
- Mission Plan: sending and updating the emergency management plans which are defined based on country's organizational structure to deal with the emergency situation.
- Permission: asking permission from an upper level organization to perform some specific action.
- Resource Management: querying, allocating, reserving and releasing resources such as fire trucks, boats, ambulances, temporary rest centers, volunteers etc.
- Scheduling: scheduling the execution of missions defined in the mission plan.
- Sensor Management: configuring sensors to gain observations from areas of interest.

- Sensor Measurement: transmitting the sensor measurements to C2 systems.
- Situation Analysis: providing simulation of the situation on the ground for a given set of data.
- Situation Reporting: transmitting timely available situation reports.
- Tracking of Citizens: tracking victims and patients during evacuation and transfer operations.
- User Authentication and Authorization: managing user authentication and providing users with the convenience and speed of a single sign-on.

3 GITB Project and the Testing Methodology

GITB project aims at developing a global testing framework, architecture and methodologies for state-of-the-art eBusiness Specifications and profiles covering all layers of the interoperability stack (organizational, semantic and technical interoperability) [4]. It promotes the reusability of testing resources and capabilities among different domains and standards. The work on GITB is motivated by the increasing need to support testing of eBusiness scenarios as a means of fostering standards adoption, achieving better compliance to standards and greater interoperability within and across the various industry, governmental and public sectors [10].

The GITB Testing Framework outlines a methodology (what should be tested, how the testing should happen, where the testing should be realized) and defines the actors and the concepts (such as the levels of testing, the difference between a Test Bed and a testing application, the test artefacts) in a testing domain. It defines the architecture as shown in Fig. 1 for modular and interoperable Test Beds, with detailed description of their components and services. Furthermore, the schemas for the definition and execution of testing artefacts such as Test Scenarios, Test Presentation and Test Reporting are also provided. The architecture also contains a global Testing Resource Registry and Repository, where any testing application and/or testing resource can be published to the users, who need any type of testing.

According to GITB testing methodology, systems are certified in two testing steps: conformance testing and interoperability testing. Conformance testing involves verifying whether a single implementation conforms to the underlying specifications. The testing is about checking the conformance of message contents to a document standard specified in the corresponding profile. Interoperability testing, on the other hand, includes more than one system, each playing different role in a given scenario. It is about testing the systems' ability to exchange information between each other by complying with the specifications defined in the profiles and to use the information that has been exchanged.

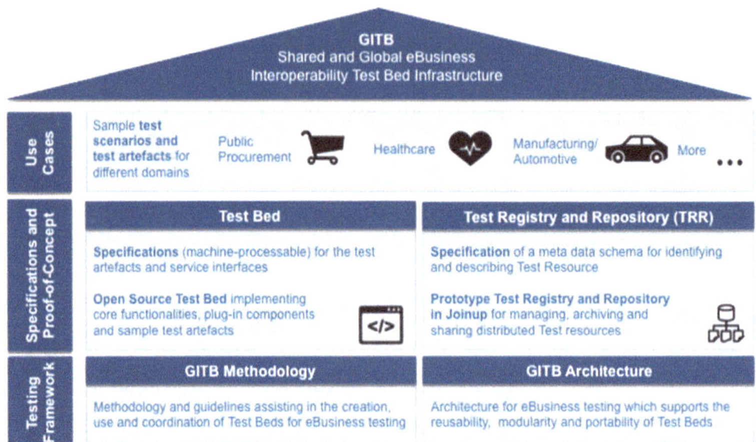

Fig. 1. GITB architecture

4 Conformance Testing

Conformance Testing is defined as verifying an artifact (e.g., an EDXL SitRep message) against the rules defined in the specification. Interactive Conformance Testing involves direct interaction between Test Bed and System Under Test (SUT), combined with dynamic validation of SUT outputs (document validation). The document validation is delegated by the Test Suite engine to a Document Validator (Fig. 2).

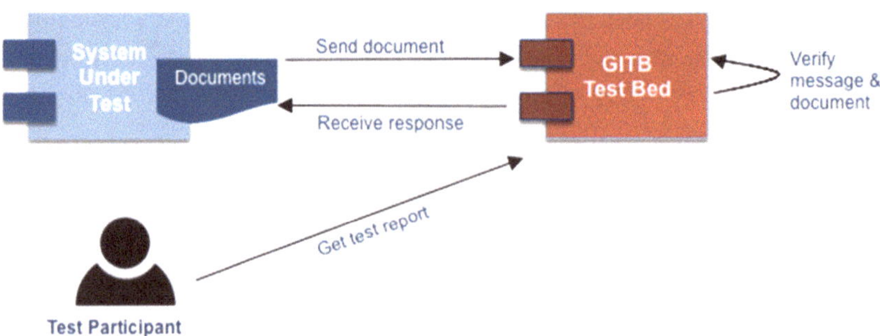

Fig. 2. Conformance testing mechanism

In C2-SENSE, conformance testing is realized through a collection of test cases each evaluating whether the captured messages satisfy the requirements, that is document validation. It includes only one system to test, which is System Under Test (SUT), and the other roles are played by GITB according to the test scenarios.

The Test Cases (scenarios) are the main drivers of the tests in GITB Framework. The test cases are described in terms of Test Description Language (TDL), which is an XML based language suggested by GITB. In this respect, when the users execute test cases, their corresponding TDL definitions are run by GITB Engine. The TDL is basically composed of XML constructs for sending/receiving messages (that are bound to messaging adapters), performing document validation and sequential/conditional/loop steps.

Table 1. Conformance test case definition of "Send Alert" message

```
<?xml version="1.0" encoding="UTF-8"?>
<testcase id="Alert_SendAlert_AlertSender" xmlns="http://www.gitb.com/tdl/v1/"
xmlns:gitb="http://www.gitb.com/core/v1/">
<metadata>
        <gitb:name>Alert_SendAlert_AlertSender</gitb:name>
        <gitb:type>CONFORMANCE</gitb:type>
        <gitb:version>0.1</gitb:version>
        <gitb:description>test whether The Alert Sender creates an alert message and sends it to Alert
Receiver.</gitb:description>
</metadata>
<namespaces>
        <ns prefix="m">urn:oasis:names:tc:emergency:EDXL:DE:1.0</ns>
        <ns prefix="cap">urn:oasis:names:tc:emergency:cap:1.2</ns>
</namespaces>
<imports>
        <artifact type="schema" encoding="UTF-8" name="CAP_XSD_File" >alert/artifacts/CAP-
v1.2.xsd</artifact>
        <artifact type="schema" encoding="UTF-8" name="EDXL-DE_XSD_File">alert/artifacts/EDXL-
DE.xsd</artifact>
        <artifact type="object" encoding="UTF-8" name="ACK_XML_File"
>alert/artifacts/messages/SampleSOAP-ContainingDE-ContainingACK.xml</artifact>
</imports>
<actors>
        <gitb:actor id="AlertSender" name="AlertSender" role="SUT"/>
        <gitb:actor id="AlertReceiver" name="AlertReceiver" role="SIMULATED" />
</actors>
<variables>
        <var name="cap" type="object"></var>
</variables>
<steps>
        <btxn from="AlertSender" to="AlertReceiver" txnId="t1" handler="SoapMessaging"/>
        <receive id="soap_output" desc="Waiting an Alert" from="AlertSender" to="AlertReceiver" txnId="t1">
                <config name="soap.version">1.1</config>
        </receive>
        <verify handler="XSDValidator" desc="EDXL-DE_XSD Validation">
                <input name="xmldocument">$soap_output{soap_content}</input>
                <input name="xsddocument" source="$EDXL-DE_XSD_File"/>
        </verify>
        <assign to="$cap"
source="$soap_output{soap_content}">//m:EDXLDistribution/m:contentObject/m:xmlContent/m:embeddedXMLCo
ntent/cap:alert</assign>
        <verify handler="XSDValidator" desc="CAP_XSD Validation">
                <input name="xmldocument">$cap</input>
                <input name="xsddocument" source="$CAP_XSD_File"/>
        </verify>
        <send id="ack" desc="Sending Positive Acknowledgement" from="AlertReceiver" to="AlertSender"
txnId="t1">
                <input name="soap_message" source="$ACK_XML_File" />
                <config name="soap.version">1.1</config>
        </send>
        <etxn txnId="t1"/>
</steps>
</testcase>
```

An example conformance test case for a SUT sending an Alert message in OASIS CAP format through SOAP is presented in Table 1. The test case consists of three parts:

1. metadata: gives the information about the test case.
2. actors: actors in defined in this test case. In this example, Alert Sender is the system under test (SUT) and the Alert Receiver is simulated by the GITB.
3. steps: steps to be executed. In this example, an alert message is expected from the Alert Sender actor and the corresponding messaging handler identified as "Soap-Messaging" is used. After that the Alert document in the message is extracted and validated by the document validator identified with "XSDValidator".

The test cases are presented to the users graphically in GITB Web page. When the execution is started, the results of the execution are displayed to the user graphically. The presentation of the above test case is displayed in Fig. 3. As it can be seen in the figure, there are four steps:

1. An EDXL-DE XML file containing the alert message in CAP format is uploaded to the system.
2. The uploaded EDXL-DE XML file is validated against EDXL-DE XSD.
3. The CAP XML extracted from EDXL-DE XML is validated against CAP Schematron.
4. An acknowledgement is sent back to SUT.

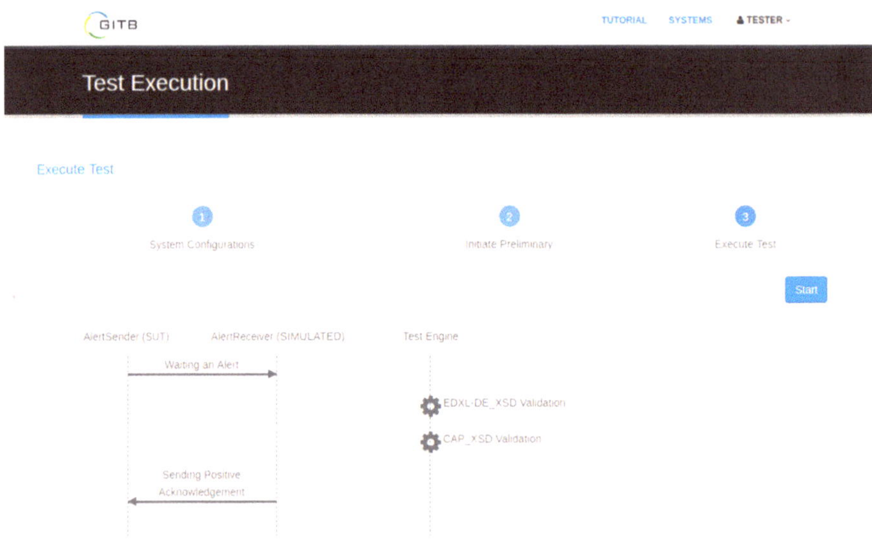

Fig. 3. Send alert test case steps

In order this system to work properly, example XML and corresponding schema files should be provided to GITB system in advance. GITB uses the XML files while imitating the actor and the schemas to validate the input files in different test cases.

Network host IP, port, and URL should also be set for SUT to communicate with GITB. When the execution starts, the system waits SUT to send an XML file containing the request. When SUT sends the file to GITB, the file is validated against its schema, acknowledgement is sent to SUT and validation results are shown to the users.

5 Interoperability Testing

Interoperability Testing is defined as the ability of two SUTs to interact with each other in compliance with the specification. This interaction usually involves data artefacts (e.g. messages) produced by one SUT and consumed by the other. Both SUTs should be configured individually and every SUT should play role of one actor in Test Suite. While actors start to communicate with each other, GITB listens the transactions between them (Fig. 4).

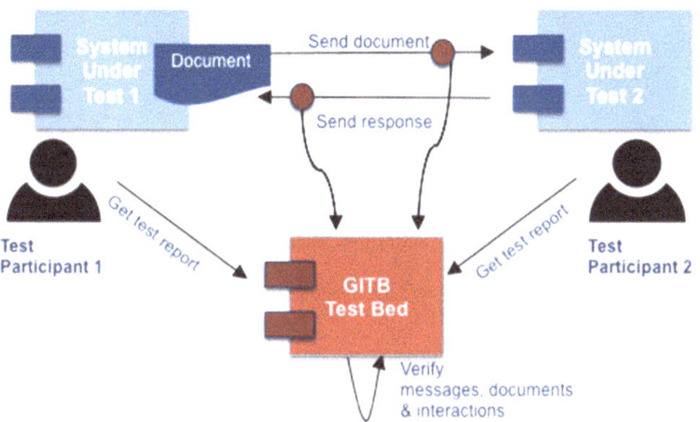

Fig. 4. Interoperability testing mechanism

In C2-SENSE, interoperability testing is realized as a process of verifying that two SUTs each playing different role in a given scenario can interoperate with each other at one or more layers of the interoperability stack, while conforming to the specifications defined in the profiles. This type of testing is executed by operating SUTs and capturing their exchanges.

An example interoperability test case for two SUTs to communicate with each other for the exchange of an Alert message through SOAP is presented in Table 2.

Table 2. Interoperability test case definition for the exchange of an alert message

```
<?xml version="1.0" encoding="UTF-8"?>
<?xml version="1.0" encoding="UTF-8"?>
<testcase id="Alert_Interoperability_SendAlert" xmlns="http://www.gitb.com/tdl/v1/"
xmlns:gitb="http://www.gitb.com/core/v1/">
<metadata>
        <gitb:name>Alert_Interoperability_SendAlert</gitb:name>
  <gitb:type>INTEROPERABILITY</gitb:type>
        <gitb:version>0.1</gitb:version>
        <gitb:description>test whether Alert Sender and Alert Receiver can interoperate</gitb:description>
</metadata>
<namespaces>
        <ns prefix="m">urn:oasis:names:tc:emergency:EDXL:DE:1.0</ns>
  <ns prefix="cap">urn:oasis:names:tc:emergency:cap:1.2</ns>
</namespaces>
<imports>
  <artifact type="schema" encoding="UTF-8" name="CAP_XSD_File">alert/artifacts/CAP-v1.2.xsd</artifact>
  <artifact type="schema" encoding="UTF-8" name="EDXL-DE_XSD_File">alert/artifacts/EDXL-
DE.xsd</artifact>
</imports>
<actors>
        <gitb:actor id="AlertSender" name="AlertSender" role="SUT"/>
  <gitb:actor id="AlertReceiver" name="AlertReceiver" role="SUT" />
</actors>
<variables>
        <var name="cap" type="object"></var>
</variables>
<steps>
        <btxn from="AlertSender" to="AlertReceiver" txnId="t1" handler="SoapMessaging"/>
        <listen id="soap_output" desc="Send message to Alert Receiver" from="AlertSender" to="AlertReceiver"
txnId="t1">
                <config name="soap.version">1.1</config>
        </listen>
        <assign to="$cap"
source="$soap_output{soap_content}">//m:EDXLDistribution/m:contentObject/m:xmlContent/m:embeddedXMLCo
ntent/cap:alert</assign>
        <listen id="ack" desc="Alert Receiver returns acknowledgement" from="AlertReceiver"
to="AlertSender" txnId="t1">
        <config name="soap.version">1.1</config>
                <config name="http.uri.extension">/axis2/services/Distribution</config>
        </listen>
        <etxn txnId="t1"/>
        <verify handler="XSDValidator" desc="EDXL-DE_XSD Validation">
        <input name="xmldocument">$soap_output{soap_content}</input>
    <input name="xsddocument" source="$EDXL-DE_XSD_File"/>
        </verify>
  <verify handler="XSDValidator" desc="CAP_XSD Validation">
        <input name="xmldocument">$cap</input>
        <input name="xsddocument" source="$CAP_XSD_File"/>
        </verify>
        <verify handler="XSDValidator" desc="EDXL-DE_XSD Acknowledgement Validation">
   <input name="xmldocument">$ack{soap_content}</input>
   <input name="xsddocument" source="$EDXL-DE_XSD_File"/>
  </verify>
  </steps>
  </testcase>
```

The test case described in TDL for interoperability testing is not much different than the one described for conformance testing. The differences in each part are as follows:

1. metadata: type is changed from conformance to interoperability.
2. actors: both actors are SUT now, GITB does not simulate any of it.
3. steps: all the messages are now validated by corresponding document validator. GITB does not send any message as part of a simulation. Since there is no simulation, the number of steps is increased.

The presentation of the test case is displayed in Fig. 5. In interoperability test case execution, the system first waits both SUTs to complete their configurations. In the configuration phase, each SUT chooses the actor that they will be acting in the scenario (e.g. Alert Sender or Alert Receiver). When the execution starts, receiver SUT waits the sender SUT to send a message. Sender SUT sends the message to GITB, GITB validates the message and forwards it to receiver SUT. After message is retrieved by the receiver SUT, it sends the acknowledgement message to GITB, GITB validates it and forwards to sender SUT. If validation fails at any step or a SUT does not send the desired message, GITB shows a detailed error report.

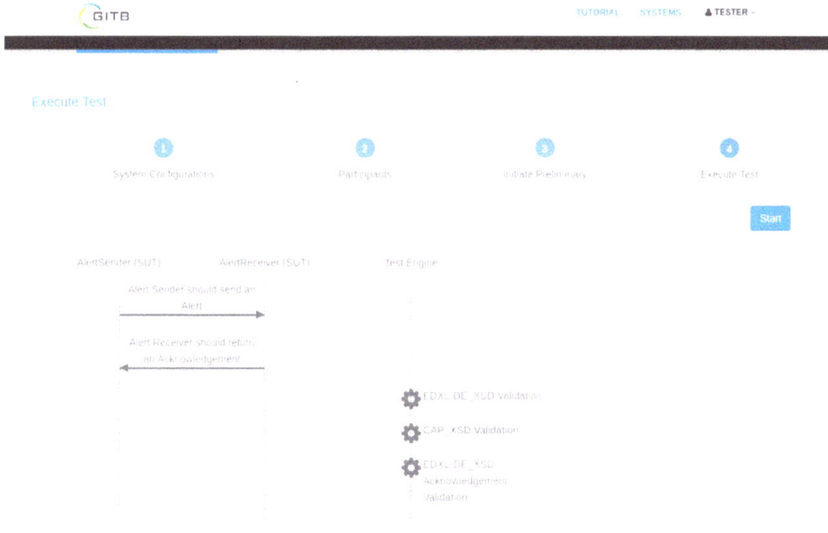

Fig. 5. Alert message exchange test steps

6 Conclusion

The correct information exchange among applications can only be guaranteed and systems can be certified through conformance and interoperability testing. Conformance testing involves checking whether the applications conform to the standards, while interoperability testing involves checking whether they can interoperate with each other in compliance with the specification. In this paper, we described a certification mechanism implemented for emergency management domain based on the principles and specifications defined in GITB initiative in the scope of C2-SENSE project.

In C2-SENSE, a number of Emergency Interoperability Profiles have been defined and an Emergency Interoperability Framework has been implemented based on these profiles. After the implementation, the conformance of the applications existing in the framework to the Emergency Interoperability Profiles and their interoperability have been tested by the mechanism described in this paper, and the framework has been certified.

Acknowledgements. The work presented in this paper is achieved in the scope of C2-SENSE project [5] supported by the European Community's Seventh Framework Programme (FP7/2007-2013) under grant agreement number 607729.

References

1. Sommerville, I.: Software engineering, pp. 41–465. Addison-Wesley, Boston (2011)
2. Namli, T., Aluc, G., Sinaci, A.A., Kose, I., Akpinar, N., Gurel, M., Arslan, Y., Ozer, H., Yurt, N., Kirici, S., Sabur, E., Ozcam, A., Dogac, A.: Testing the conformance and interoperability of NHIS to Turkey's HL7 profile. In: The Proeeding. of 9th International HL7 Interoperability Conference 2008, Crete, Greece, October 2008
3. Namli, T., Dogac, A., Sinaci, A.A., Aluc, G.: Testing the interoperability and conformance of UBL/NES based applications. In: The Proceeding of eChallenges Conference, Istanbul, Turkey, October 2009
4. GITB Project - Global eBusiness Interoperability Testbed methodologies. https://joinup.ec.europa.eu/asset/cen-ws-gitb/description. Accessed 16 May 2017
5. C2-SENSE Project - Interoperability Profiles for Command/Control Systems and Sensor Systems in Emergency Management. http://c2-sense.eu/. Accessed 16 May 2017
6. Tolk, A.: Beyond technical interoperability-introducing a reference model for measures of merit for coalition interoperability. Old Dominion University, Norfolk, VA (2003)
7. Gençtürk, M., Arisi, R., Toscano, L., Kabak, Y., Di Ciano, M., Palmitessa, A.: Profiling approach for the interoperability of command & control systems with sensing systems in emergency management. In: Proceedings of the 6th Workshop on Enterprise Interoperability, Nîmes, France (2015)
8. Božić, B., Gençtürk, M., Duro, R., Kabak, Y., Schimak, G.: Requirements engineering for semantic sensor in crisis and disaster management. In: Environmental Software Systems. Infrastructures, Services and Applications, pp. 397–406. Springer International Publishing, New York City (2015)
9. Gencturk, M., Duro, R., Kabak, Y., Božic, B., Kahveci, K., Yilmaz, B.: Interoperability profiles for disaster management and maritime surveillance. In: eChallenges e-2015 Conference, 2015, pp. 1–9. IEEE, November 2015
10. Dogac, A.: Draft CEN Workshop Agreement: Global eBusiness Interoperability Test Bed (GITB) Phase 3: Implementation Specifications and Proof-of-Concept, p. 9 (2015). Accessed 17 Apr 2017

Civil Protection Organisational and Procedural Interoperability Profile

Jan Piwiński[1(✉)], Biagio Lanziani[2], Francesco Ronco[3], and Ivana Caputo[3]

[1] Industrial Research Institute for Automation and Measurements,
Al. Jerozolimskie 202, 02-486 Warsaw, Poland
jpiwinski@piap.pl

[2] Regola s.r.l., c.so Turati 15/H, 10128 Turin, Italy

[3] Sezione Protezione Civile - Regione Puglia,
Via Delle Magnolie 6/8, 70026 Modugno Z.I., BA, Italy

Abstract. Paper presents the results of the results of the C2-SENSE activities in project's Work Package Organisational Interoperability, where we provided template for writing of the Organisational Interoperability Profiles (OIPs) as well as the framework for assessing the existing organisational agreements that is aligned with this template and tested by analysing several existing agreements. OIPs cover Aligned Operations and Procedures and Harmonized Strategy/ Doctrines layers of the interoperability stack. These layers address the cooperation among organisations at procedural or operational level.

C2-SENSE Organisational Interoperability Profile template enables any organisation involved in the crisis management to prepare and write new inter-organisation agreements, which describe the nature of cooperation in everyday operations among them. The template is aligned with relevant international standards recommendations and applicable to all types and sizes of organisations that wish to prepare, maintain or improve their inter-organisation partnering agreement and demonstrate conformity to other organisation, when arranging the initial dialogue with potential partners.

The standards used in preparing the C2-SENSE OIPs address the organisational structures and internal procedures of the involved parties. Resulting OIPs, therefore, ensure a degree of interoperability among organisations in planning, establishing, implementing, operating, monitoring, reviewing, maintaining operations, even in the situation, when their organisations structures or internal procedures are different. This can help emergency organisations to recognize and resolve their cross-organisational interoperability issues.

Keywords: Profile · ISO standards · Organisations · Agreement

1 Introduction

A wide range of threats and hazards can result in destabilising or disruptive events and escalate towards unpredictable and large scale consequences for Societal and Citizen

R. Szewczyk and D. Havlik (eds.), *Recent Trends in Control and Sensor Systems in Emergency Management*, Advances in Intelligent Systems and Computing 675,
https://doi.org/10.1007/978-3-319-70452-4_9

Security. Both public and private stakeholders require adequate solutions in organisation, procedures and technological capabilities to respond effectively.

Societal and Citizen Security is dedicated to enabling and improving the capability of public and private stakeholders to prepare for, respond to and recover from such destabilizing or disruptive events.

Various private and public organisations have organisational and technological capabilities to prevent incidents and to cope with the consequences. Thus, there is a need to develop specific standards to enable these organisations, on local, regional, federal, national and European level to be effectively coordinated and to cooperate with other organisations before, during and after a destabilising or disruptive event.

The development of European standardisation activities within the concepts of Societal and Citizen Security will permit both public and private stakeholders to acquire a common approach for all relevant issues, e.g. human, organisational, technical and functional interoperability, management of destabilising or disruptive events and business continuity capabilities [1].

Standards play a major role in defragmenting markets and helping industry in achieving economies of scale. Standards are also of upmost importance for the demand side, notably with regard to interoperability of technologies used by first responders, law enforcement authorities, etc. Additionally, standards are essential for ensuring uniform quality in the provision of security services. Creating EU-wide standards and promoting them on a worldwide level is also a vital component of the global competitiveness of the EU security industry.

However, few EU-wide standards exist in the security area. Divergent national standards seem to pose a major obstacle for the creation of a true internal market for security, thus hindering the competitiveness of EU industry [2].

While interoperability has been discussed conceptually in the information systems (IS) literature, few comprehensive empirical studies have been conducted to conceptualize this construct and examine it in depth. For instance, it is unclear how interoperability is formed and whether it can improve organisational performance. To fill the gap, we argue that inter-organisational systems (IOS) standards are a key information technology infrastructure facilitating formation of interoperability. As an organisational ability to work with external partners, interoperability's development depends not only on capability building within firm boundaries but also on community readiness across firm boundaries.

Furthermore, our results show that interoperability acts as a mediator by enabling firms to achieve performance gains from IOS standards adoption. Our study sheds new light on formation mechanisms as well as the business value of interoperability [3].

Nowadays interoperability is possible only when a common language is used by various IS, despite heterogeneity in software, hardware, and system architecture. Such a common language is defined by IOS standards, technical specifications describing data formats, and communication protocols for computer communications [4, 5]. IOS standards contribute to interoperability by providing "shared business terms, functions, processes, and protocols". Companies need to carefully manage inter-organisational processes in order to access external resources, mitigate strategic uncertainties, and gain competitive advantage.

Despite the critical role of interoperability, theoretical and empirical research pertaining to this important organisational capability is limited in the IS field. Our literature review indicates that interoperability has never been formally examined in prior empirical studies of inter-organisational systems [6, 7].

2 Scope of the Profile

The Organisational Interoperability Profiles describe and illustrate the methodology that allows the end-user organisations to prepare their partnering agreements or organisational procedures in formalized way.

Organisational Interoperability Profiles supports organisations responsible for civil protection in preparing and writing new inter-organisation agreement, which describe the nature of cooperation in everyday operations among them.

The Organisational profile for establishing partnering agreement, is aligned several international standards and provide a guideline on how to prepare and establish inter-organisation partnering agreement, which can be used as a template for formal documentation of such cooperation agreements for organisations involved in civil protection mechanism.

The organisational profiles that are generic and not specific for a country or organisation and they are applicable to all types and sizes of organisations that wish to:

- Prepare, establish, maintain and improve their partnering agreement,
- Demonstrate conformity to other organisation, when arranging the initial dialogue with potential partners,
- Make a self-determination and self-declaration of conformity with this International Standard, when preparing their partnering agreement.

The research concentrates also on analysis of existing international standards, encapsulating the relevant material from their recommendations and also focuses on analysis of existing organisational agreements among civil protection sector.

The results of this work will be used to demonstrate how misalignments in the procedures and agreements of different organisations can be recognised according to relevant standard.

The OIP is the result of an analysis and comparison of the existing formal (European and international) security standards implemented in Europe. An organisational profile was developed on the basis of needs and requirements analysis of following Standards, which was the main examined material during preparation of this proposal:

- CEN TC 391 - Societal and Citizen Security [1, 8]
- Mandate M/487 to Establish Security Standards. Proposed standardization work programmes and road maps [2]
- ISO 22397:2014 - Societal security — Guidelines for establishing partnering arrangements [9]
- ISO 22301:2012 - Societal security — Business continuity management systems — Requirements [10]
- ISO 22300, Societal security — Terminology [11]

CEN, CENELEC and ETSI are the official providers of European Standards and technical specifications. Their activities are set out by the Regulation 1025/2013 for the planning, drafting and adoption of European Standards and other deliverables in all areas of economic activity.

3 Organisational Profile for Establishing Partnering Agreement

The "Organisational profile for establishing partnering agreement" provides a guideline on how to prepare and establish inter-organisation partnering agreement, which we believe can be used as a template for formal documentation of such cooperation agreements for organisations involved in civil protection mechanism.

This profile is aligned with aforementioned Standards and the material for preparation of this profile was encapsulated from analysed international standards, therefore we claim that it is applicable to all types and sizes of organisations.

The organisation shall determine external and internal issues that are relevant to its purpose and that affect its ability to achieve the intended outcome(s) of inter-organisation partnering agreement. This profile covers the following components:

- Item 1 - **Preparation actions** - it summarises the genesis of partnering agreement, including suggestions on how to arrange initial dialog, search and identify partners to agreement.
- Item 2 - **Understanding of the organisations and their context** - It introduces requirements necessary to establish the context of the partnering agreement as it applies to the organisation, as well as needs, requirements, and scope.
- Item 3 - **People and management** - It summarizes the requirements specific to roles and relationship rules division and provide suggestions to management and responsibility role in the partnering agreement.
- Item 4 - **Suggestions and facilitators to crisis response planning** - It describes overall suggestions to establishing strategic objectives and guiding principles for the partnering agreement in crisis response domain stemming from Communication (Semantics), reporting and warning systems, operational efficiency, Training and testing.
- Item 5 - **Law** - It supports partnering agreement operations while documenting, controlling, maintaining and retaining required documentation.

The chapters below present the selected findings from Item 4 Suggestions and facilitators to crisis response planning

Suggestions and Facilitators to Communication (Guidance on Semantics)

To better understand and distinguish between different concepts and facilitate communication and understanding (before, during and after crises) partners shall agree to common terminology or common language of work and generic models. Therefore partners shall generate a dictionary comprising at least the most important European languages in addition to the vocabulary list ISO 22300 to facilitate communication. For example they shall provide definition of risk manager, crisis, crisis room, emergency, resilience. When developing the partnering agreement partners shall:

- Develop standardised common geospatial basic information (based on existing GIS standards) to be used by organisations before and during crisis situations (for these organisations to provide additional information to the common base or to retrieve information to be consolidated within their own systems). This common geospatial basic information should use minimum semantic agreements and minimum standardised icons.
- Facilitate radio communication interoperability (voice, data, image) to develop standards for the usage of mobile broadband services in addition to the professional mobile radio PMR. This shall improve the information exchange of emergency management organisations (e.g. based on LTE, WIFI, whatsoever).
- Develop an easy and standardized way to link arbitrary smart phones together in order to exchange incident-relevant data.
- To assist first responders localisation include Geo-localization (GIS) standards for use in buildings and underground systems to facilitate FR intervention. It is needed for providing dynamic information during an emergency (i.e. evacuation information in real time, location, infrastructure availability, exit routes availability).
- Develop communication interoperability by a better definition of needs and the use of minimum common semantic and minimum set of requirements. It shall be implemented on a volunteer basis, considering existing implementations. This would eventually allow progressive standardization of event description and of digital objects, adaptation to evolving technologies and facilitate mechanisms to share information on a day-by-day basis.

Guidance on Reporting and Warning Systems
When developing the partnering arrangement the organisation shall establish, implement and maintain procedures for

- Incident management in first hours to develop client-based applications to decode alert messages in consumer receivers (smart phone, tablet, etc.) for specification use of navigation enabled devices for alerting and establishing a standard way to refer to administrative areas with geo-codes that are valid all over Europe for alerting purposes. For more details look at comments: consider ISO 22324 "colour-coded alert" [12].
- Reinforce citizens and local territorial community awareness and involvement by:
- Increase knowledge of risks and available channels for information and advice for appropriate actions (before, during and after the incident).
- Warning (alert and notification) dissemination understanding.

- Develop alert libraries that are applicable in all European countries. Define common European messages schemes for fire and evacuation systems. For more details use ISO 22322 "Public warning" [13] defined process.
- Standardization of the way of acquiring digital information from victims/public and sending it to the whole command & control system (it may include developing a common 'victim ticket', to be filled in by victims using smart phone emergency applications). Consider issues such as the protection of personal information or the impact of national legislation or the saturation of the Public Safety Answering Point (PSAP).
- Development alert libraries that are applicable in all European countries: Develop a communication protocol that allows lightweight transmission of alert messages and supports light encoding of the alert libraries, with possible use of wireless media (suggest more specific use of CAP, based on alert libraries, to allow interoperability).
- Usage of social media for early detection through weak signals and establish methodology for sourcing information (social media, tweets, crowd source information) to assess impact of wide scale disaster and identify public needs. Secondly develop a common and standardized procedure in order to let citizens actively bring in their resources into the relieve effort.

Guidance on Operational Efficiency
The organisation shall establish, document, and implement procedures and management structure to respond to a disruptive incident using personnel with the necessary responsibility, authority and competence to manage an incident, therefore when establishing partnering agreement organisations shall:

- Define "limited key information" to share (pre, during, post incident) to improve preparedness, coordination and debriefing (between different actors and different hierarchical levels).
- Develop methodologies for anticipation and decision making process under uncertainty (when there is a lack of information, unreliable situation assessment, uncertainty about situation evolution).
- Define exercises evaluation procedures: Crisis Management performance parameters, identified gaps, communication/planning/implementations of findings, develop lessons learned data base, produce a common lessons identified process (identification, implementation, inclusion in Standard Operating Procedures SOPs or training courses). For more details look at ISO 22398 guidelines for exercises [14].
- Enhance the assistance for victims management organisations shall establish standards on patient-management in mass casualty incidents (e.g. minimal data-set for patient-management in mass casualty incidents, management of data of affected persons in mass casualties). To close the gap in (inter)national pre-hospital patient-management with differing national standards. Develop a standardized electronic triage system to improve the logistics and the situation awareness.
- Improve the management of vertical bottom-up information flow for situation assessment, both within the public sector and within private organisations to facilitate and accelerate real understanding of key issues, critical information, priorities and to

develop capacity to anticipate situation evolution by a better understanding of next layer expectations.

Training and Testing
The organisation shall exercise and test its procedures to ensure that they are consistent with its partnering agreement objectives. The partnering agreement shall explicitly describe:

- The training procedures on how to run simple exercises (plan, execute and report).
- Involvement of citizens, communities and organisations with plans to increase community resilience.
- Pan-European collective training (table-top, simulation, operational).
- Multi-agency, common cross-border training program (share best practices, networking, get to know each other, continuous improvement). For more details please check developments from ISO 22398 (guidelines for exercises).

4 Conclusions

This paper presents Organisational Interoperability Profiles (OIPs), that shall increase the harmonisation of the European security market and reduce fragmentation with the establishment of a set of comprehensive European standards and enhance secure interoperable communications and data management between the various security control centres, operators, public authorities and first responders.

Due to this fact that there are so many misalignments among international security procedures and frameworks, in situation when two organisations from two different countries want to cooperate on operational level, for example for multinational forest firefight trainings actions, we recommend to use "Organisational profile for establishing partnering agreement", which provides a guideline on how to prepare and establish inter-organisation partnering agreement in civil protection domain.

The research presented in this paper can be treated as an understandable roadmap of existing standards, which can help public authorities to understand usage of standards and also can help standardization developers to provide useful standards, also can get major stakeholders and Public Authorities to understand the use of standards and apply them. The incorporation of the aforementioned recommendations will ensure that the defined organisational profiles are in line with the stakeholders' needs and requirements.

The set of guidelines and recommendations defined in the document are designed to enhance operational interoperability for emergency service and civil protection agencies when utilising C2-SENSE system in crisis situations. Indeed, it is important to highlight that these guidelines and procedures do not intend to replace, nor bypass, the daily arrangements of each Responder Agency's operation, but to improve and strengthen inter-agency collaboration.

References

1. CEN/TC 391 Business Plan. Revision 02, 23 Nov 2010
2. European Commission: Mandate M/487 to Establish Security Standards. Final Report Phase 2. Proposed standardization work programmes and road maps, 05 July 2013
3. Zhao, K., Xia, M.: Forming interoperability through interorganisational systems standards. J. Manag. Inf. Syst. Spring **30**(4), 269–298 (2014)
4. Nelson, M., Shaw, M.J., Qualls, W.: Interorganisational system standards development in vertical industries. Electron. Mark. **15**(4), 378–389 (2005)
5. Zhu, K., Kraemer, K.L., Gurbaxani, V., Xu, S.X.: Migration to open-standards interorganisational systems: network effects, switching costs, and path dependency. MIS Q. **30**, 515–539 (2006). special issue
6. Rai, A., Tang, X.: Leveraging IT capabilities and competitive process capabilities for the management of interorganisational relationship portfolios. Inf. Syst. Res. **21**(3), 516–542 (2010)
7. Venkatesh, V., Bala, H.: Adoption and impacts of interorganisational business process standards: role of partnering synergy. Inf. Syst. Res. **23**(4), 1131–1157 (2012)
8. https://standards.cen.eu/dyn/www/f?p=204:7:0::::FSP_ORG_ID:680331&cs=18422BF6F2CD25C72E8F633D87A8147AB
9. http://www.isotc292online.org/publications/iso22397/
10. https://www.iso.org/obp/ui/#iso:std:iso:22301:ed-1:v2:en
11. https://www.iso.org/obp/ui/#iso:std:iso:22300:dis:ed-2:v1:en
12. https://www.iso.org/standard/50061.html
13. https://www.iso.org/standard/53335.html
14. https://www.iso.org/standard/50294.html

Author Index

R. Szewczyk and D. Havlik (eds.), *Recent Trends in Control and Sensor Systems in Emergency Management*, Advances in Intelligent Systems and Computing 675,
https://doi.org/10.1007/978-3-319-70452-4